Systems 4.0

The underlying premise for Industry 4.0 is a systems approach. This book introduces the concept of Systems 4.0 as a foundational requirement for the success of Industry 4.0 in the same way that Quality 4.0 has emerged to advance Industry 4.0.

Systems 4.0: Systems Foundations for Industry 4.0 discusses the role of the professional engineer in advancing commerce and industry. It offers an introduction to Industry 4.0 and how to leverage the digital era to improve industrial operations. The book presents and describes the first industrial revolution on through to the fourth revolution and provides general systems engineering principles that can be used with Industry 4.0.

This is a practical guide for professional engineers and consultants involved in Industrial Engineering, Mechanical Engineering, Operations Management and can also be used as a reference for students.

Analytics and Control Series

Series Editor: Adedeji B. Badiru, Air Force Institute of Technology, Dayton, Ohio, USA

Series Description: Decisions in business, industry, government, and the military are predicated on performing data analytics to generate effective and relevant decisions, which will inform appropriate control actions. The purpose of the focus series is to generate a collection of short-form books focused on analytic tools and techniques for decision-making and related control actions.

Mechanics of Project Management
Nuts and Bolts of Project Execution
Adedeji B. Badiru, S. Abidemi Badiru, and I. Adetokunboh Badiru

The Story of Industrial Engineering
The Rise from Shop-Floor Management to Modern Digital Engineering
Adedeji B. Badiru

Innovation
A Systems Approach
Adedeji B. Badiru

Project Management Essentials
Analytics for Control
Adedeji B. Badiru

Sustainability
Systems Engineering Approach to the Global Grand Challenge
Adedeji B. Badiru and Tina Agustiady

Operational Excellence in the New Digital Era
Adedeji B. Badiru and Lauralee Cromarty

Artificial Intelligence and Digital Systems Engineering
Adedeji B. Badiru

Systems 4.0
Systems Foundations for Industry 4.0
Adedeji B. Badiru, Olufemi A. Omitaomu

For more information on this series, please visit: https://www.routledge.com/Analytics-and-Control/book-series/CRCAC

Systems 4.0
Systems Foundations for Industry 4.0

Adedeji B. Badiru
Olufemi A. Omitaomu

CRC Press
Taylor & Francis Group
Boca Raton London New York

CRC Press is an imprint of the
Taylor & Francis Group, an **informa** business

First edition published 2023
by CRC Press
6000 Broken Sound Parkway NW, Suite 300, Boca Raton, FL 33487-2742

and by CRC Press
4 Park Square, Milton Park, Abingdon, Oxon, OX14 4RN

CRC Press is an imprint of Taylor & Francis Group, LLC

Library of Congress Cataloging-in-Publication Data
Names: Badiru, Adedeji Bodunde, 1952- author. | Omitaomu, Olufemi Abayomi, author.
Title: Systems 4.0 : systems foundations for industry 4.0 / Adedeji B. Badiru, Olufemi A. Omitaomu.
Description: First edition. | Boca Raton : CRC Press, 2024. | Includes bibliographical references and index.
Identifiers: LCCN 2023005580 (print) | LCCN 2023005581 (ebook) | ISBN 9781032319186 (hardback) | ISBN 9781032319902 (paperback) | ISBN 9781003312277 (ebook)
Subjects: LCSH: Industry 4.0 | Systems engineering--Technological innovations.
Classification: LCC T59.6 .B33 2024 (print) | LCC T59.6 (ebook) | DDC 670.285--dc23/eng/20230329
LC record available at https://lccn.loc.gov/2023005580
LC ebook record available at https://lccn.loc.gov/2023005581

ISBN: 978-1-032-31918-6 (hbk)
ISBN: 978-1-032-31990-2 (pbk)
ISBN: 978-1-003-31227-7 (ebk)

DOI: 10.1201/9781003312277

Typeset in Times
by SPi Technologies India Pvt Ltd (Straive)

Dedicated to the spirits of those who came before us to promote the systems pathway upon which we now tread.

Contents

Acknowledgments

We thank Cindy Carelli and her editorial team at Taylor and Francis for their extraordinary support throughout the production of this book. We also thank our families for giving us the leeway and freedom to spend countless hours from the dinner table to work on the manuscript. Now that the deal is done, we will be back at the family and fellowship dining tables.

Preface

Industry 4.0 is simply the Fourth Industrial Revolution. The common catchphrase for it is Industry 4.0. Sometimes, it is referred to as 4IR. The fundamental characteristics of Industry 4.0 is that it leverages rapid changes in technologies, the design of industries, the utilization of digital tools, technical innovation, societal preferences, and creative workforce development. Essentially, Industry 4.0 centers on the three ingredients of an organization:

- People
- Technology
- Process

How these three elements interface, interact, and interconnect in a global environment is what determines the efficacy of Industry 4.0. Since many of the pieces of these elements are already available in the nooks and corners of many organizations, albeit in often-disjointed fashion, an overarching systems-based linkage is required. It is through systems thinking and implementation that the society can fully realize the benefits of Industry 4.0, which is a feature of a 21st-century operation. Although the moniker of Industry 4.0 is new, the underlying ideas are not new. The tools and techniques are just becoming more digitally based and integrated, from a systems framework.

The Fourth Industrial Revolution was introduced officially in 2015 by a team of scientists charged with developing a high-tech strategy for the German government. It has since spread widely and fast through various international fora, including the World Economic Forum (WEF). Quality 4.0 is a direct natural outgrowth of Industry 4.0. Using a systems framework can help bridge the talent gap in digital operations, particularly where data barriers exist. Hence, the need for the proposed book on Systems 4.0.

Author Biographies

Dr. Adedeji B. Badiru is a Professor of Systems Engineering at the Air Force Institute of Technology (AFIT). He is a registered professional engineer and a fellow of the Institute of Industrial Engineers as well as a Fellow of the Nigerian Academy of Engineering. He has a BS degree in Industrial Engineering, an MS in Mathematics, an MS in Industrial Engineering from Tennessee University, and a PhD in Industrial Engineering from the University of Central Florida. He is the author of several books and technical journal articles and has received several awards and recognitions for his accomplishments. His special skills, experience, and interests center on research mentoring of faculty and graduate students.

Dr. Olufemi A. Omitaomu is a Senior R&D Staff in Computational Urban Sciences Group within the Advanced Computing Methods for Engineering Systems Section in Computational Sciences and Engineering Division at Oak Ridge National Laboratory (ORNL). He has more than 15 years of experience in research, development, and demonstration of innovative systems and methods in the areas of energy infrastructure siting, energy systems resilience, and disaster risk analysis. He has two issued US patents and several invention disclosures. He is the lead developer of major Department of Energy's technologies including OR-SAGE (Oak Ridge Siting Analysis for Generation Expansion), CoNNECT (Citizen Engagement for Energy Efficiency and Renewable Integration), and Precision Deicer. He received the 2021 R&D 100 Award for Precision Deicer and the 2015 R&D 100 Finalist Award for CoNNECT.

Fundamentals of Industry 4.0

1

INTRODUCTION

Our world itself is from a system of systems. From our planetary system to our social system, we thrive on systems frameworks that continue to shift and advance, as new knowledge and opportunities develop. The theme of this book is that Industry 4.0, in any ramifications we desire, ultimately depends on a systems framework. Systems framework is essential for the success of Industry 4.0 (Badiru, 2009, 2022). But what is Industry 4.0? The full answer to this comes in the flavor of the related topics addressed in this chapter. Basically, Industry 4.0 is the prevailing stage of our industrial revolution. It is a multi-dimensional operational framework within our modern composition of people, process, and technology. Since technology is changing and evolving rapidly, the concept of Industry XX.0 is also evolving rapidly. We are already seeing the initial rumblings of Industry 5.0 (Elangovan, 2022). Let us focus on Industry 4.0 now so that we will have a full understanding to prepare us for Industry 5.0, when it actually arrives. The transition point will not be brusque. It is expected to be a phased-in transition with overlapping practices, tools, and techniques.

ELEMENTS OF INDUSTRY 4.0

Industry 4.0 is simply another name for the 4th Industrial Revolution. Thus, Industry 4.0 equates to 4th Industrial Revolution, although many practitioners dispute the one-to-one correspondence, for a good reason. Industry

DOI: 10.1201/9781003312277-1

4.0 is the latest possible approach to manufacturing and its value chain. It leverages the latest technologies, including smart systems, data analytics, process automation, machine learning, cloud computing, and so on. In essence, Industry 4.0 leverages the latest tools and techniques available to industry. Technology adaption and adoption are the hallmarks of Industry 4.0. Industry 4.0 integrates digital and physical systems with mainstream automation, traditional, and next-generation technologies. The mandate for integration is what makes Industry 4.0 attractive to manufacturers. This is why this book has the premise of a systems approach, which leverages the framework offered by the trademarked DEJI Systems Model® for Design, Evaluation, Justification, and Integration (Badiru, 2023). This is covered in a subsequent chapter. The integration components of the DEJI Systems Model facilitates real-time collaboration between production facilities and concomitant operations, such as research, design, assembly, production, packaging, marketing, shipping, and so on. Further, it permits the incorporation of the people aspects of production. This includes employees, external partners, vendors, sponsors, investors, and others. A good Industry 4.0 will also have a good personnel management system. Under Industry 4.0, the manufacturer is adept and adaptive to customized production and personalized human factors and ergonomics. The cloud-computing connectivity of Industry 4.0 makes it possible to craft and execute contemporary organizational strategies. Cretu (2020) presents five important features of Industry 4.0:

1. New and fully automated approach to manufacturing production, supply chain, and life cycle management
2. Charged by smart systems, real-time data, and machine learning
3. Connects digital and physical systems with mainstream automation
4. Supports instant collaboration between factory divisions
5. Helps you define your factory's growth strategy and transform the way your factory operates

The features provide a foundation for a competitive edge for a manufacturer. Of course, if all manufacturers are dancing to the same tune of Industry 4.0, who then has the competitive edge? The difference is often in terms of the project management execution processes utilized by a manufacturer. In this context, this book advocates not only the technology sides of Industry 4.0, but also the project management essentials for executing Industry 4.0 pursuits. With proper project management (Badiru, 2019), the pursuits of Industry 4.0 become more realizable. For this reason, this book also includes a chapter on the general fundamentals of project management.

The term "Industry 4.0" was coined in 2011 at Hannover Messe, a German international platform and a hotspot for industrial transformation. It was released as a high-tech strategy project in an effort to define the digitization strategy of the German Industry and model the 4th Industrial Revolution. The research had to achieve a smart factory vision by combining factory automation not only with data and machine intelligence, but also with high-quality services, custom products and intelligent monitoring. The end of the Hannover Messe project was to achieve a fully-automated approach to manufacturing and its entire value chain. Thus, the birth of Industry 4.0. In terms of worldwide terminologies, German manufacturers prefer Industry 4.0 (no surprise). The US and European countries prefer IoT (Internet of Things) and similarly-coined or concocted terms. For example, we have UOI (Universal Object Interaction) and POI (Programmable Object Interfaces). No doubt, many organizations come up with their own unique terminologies to describe the essence of Industry 4.0. The bottom line is that Industry 4.0 revolutionizes manufacturing in a way similar to how the Industrial Revolution revolutionizes industry in serial pathways.

- The First Industrial Revolution introduced steam power to mechanize production.
- The Second Industrial Revolution introduced electric power to create mass production.
- The Third Industrial Revolution introduced electronics and Information Technology to automate production.
- The Fourth Industrial Revolution introduced the internet and smart machines to connect the entire value chain of manufacturing.

What will they think of next?
Waiting in the wings is Industry 5.0 (see Elangovan, 2022).

The Smart Factory concept is when factory adopts the Cyber-Physical Systems or CPS (combination of physical technology and cyber technology) and aggressively integrates previously-independent separate systems, making the automation technologies more complex and precise (Zaidin et al., 2018). IoT or, to use the full Industry 4.0 term, Industrial Internet of Things (IIoT), in general, is a way that industry uses digital systems and smart electronics in their production system of a product in order to dynamically respond to a global supply chain.

With respect to the interchangeability of terminologies in a chronological sense, Industry 4.0 was developed (or introduced) after the emergence of the fourth industrial revolution. It was supposed to (or expected to) help shape the emerging fourth industrial revolution. Industry 4.0 unifies known and existing technologies, such as cloud computing, IoT, and cyber-physical systems, with newer technologies, such as big data, direct digital manufacturing (3D printing), and artificial intelligence (AI). Consequently, Industry 4.0 means the fourth Industrial Revolution in the manufacturing sector and its attendant value chain. It connects and integrates everything, in accordance with the tenets of systems integration. Again, this is what informs the focus of this book with respect to Systems 4.0. Systems 4.0 is introduced as a driver for the pursuit and accomplishment of Industry 4.0. With a systems-based approach, many things become more possible. Among the challenges of Industry 4.0 that a systems approach can ameliorate are the following:

- How to improve decision-making processes.
- How to enhance a team's productivity.
- How to streamline operation process.
- How to reduce costs.
- How to deliver products faster and with better quality.

In today's complex manufacturing environment, addressing the above challenges must leverage the next-generation technologies. Therein lies the appeal and necessity of Industry 4.0. Where there is smoke, there is fire. Where there are challenges, there are benefits. Implementing Industry 4.0 technologies in manufacturing has the following benefits (Cretu, 2020; Badiru, 2020):

- Industry 4.0 facilitates operational connectivity.
- Industry 4.0 paves the way for market-share growth, the evolution of operations, and the incorporation of innovation.
- Industry 4.0 leverages smart manufacturing process. Machines easily communicate with one another, collect and analyze live data, exchange insights and make smart decisions either with or without minimal human supervision. Employees can focus on high-value tasks.
- Industry 4.0 facilitates fast, data-driven business decisions leading to streamlining of operations, maximization of capacity, extension of equipment lifespan, and the improvement of production and quality.
- Industry 4.0 facilitates custom products and optimizes the supply chain.
- Industry 4.0 can increase employee productivity and boost workforce retention while cutting costs.

- Industry 4.0 technologies introduce operational excellence, transparency, flexibility, and efficiency of the production processes, infuse supply chains with data-driven knowledge, deliver positive customer experiences.

TECHNOLOGIES FOR INDUSTRY 4.0

A technology summary provided by Cretu (2020) suggests the following technologies pertinent for Industry 4.0.

Cloud computing: Industry 4.0 is all about flexibility. Adopting cloud computing fits the need for flexibility. It provides the perfect opportunity to scale infrastructure based on specific needs. In the applications area, cloud computing facilitates cost and overhead reduction by reducing the necessity to update and maintain local and independent servers. It enhances operational scalability and access to automatic updates. It improves responsiveness to workload surge through real-time shop floor access and analysis. Cloud computing provides the basis for on-demand delivery, streamlined collaboration, and transparency between business units and key stakeholders. Further, cloud computing improves risk management and business continuity in case of unexpected failures, through data backup in a safe and secure environment.

3D Printing: 3D printing, also known as direct digital manufacturing, offers production flexibility, avoiding the downside of traditional manufacturing by accommodating fast improvements and new product releases. Applications include rapid prototyping, factory performance, production time save and cost reduction in research and development.

Virtual Reality and Augmented Reality: This supports design, maintenance, assembly, training and rapid prototyping. Applications include process improvement, design and product optimization, real-time decision support.

IoT (Internet of Things): This enhances cloud connectivity with mission-critical data collected in real time directly from machines. Applications include production visibility for productivity increase, asset tracking for inventory management, predictive maintenance to reduce downtime (by up to 25 percent), accurate ship-to or bottom-up forecasting, process quality improvement, operational and energy efficiency improvement.

Cognitive manufacturing: This connects sensor-based data (IoT) with AI and advanced Analytics to search for patterns in newly collected data, apply analytics and make real-time, informed decisions based on what will happen in the future. Applications include performance asset management, resource and supply chain optimization, process improvement, digital twins, change management, product personalization, and sales and improvement dashboards

Cyber-physical systems: This creates a highly-adaptable Smart Factory system, integrating networked computers and computational algorithms with the physical world. Applications include automated guiding vehicles, distributed robotics, process control systems, and smart grid.

As a summary, parallel technologies that buttress the capabilities of Industry 4.0 include the following:

- Smart factory
- Smart manufacturing
- Digitization in manufacturing
- Predictive maintenance
- Computer vision
- IoT
- Robot factory
- Industrial automation
- Virtual and augmented reality
- Additive manufacturing
- 3D printing
- Digital twins

Elangovan (2022) presents the anticipated future of the industrial economy. Manufacturing organizations need to direct more attention to the emerging technologies to meet the requirements of the future industrial economy. Traditionally, the focus had been on basic business survival. In the emerging competitive manufacturing environment (Badiru et al., 2019), leveraging new technologies, similar to the ones listed above, may be the key to a sustainable survival. Considering the principles of industrial engineering (Badiru, 2014), work simplification can facilitate the ideals of Industry 4.0 and beyond. Foidl and Felderer (2016) point out the need for more contemporary research on the challenges and opportunities of Industry 4.0.

SITUATIONAL AWARENESS FOR INDUSTRY 4.0

Everything is industrially interconnected in the modern marketplace. It is a systems world and all organizations must demonstrate a systems view of the world when managing international projects. Teamwork and mutual understanding take on a different flavor when different cultures are involved. Not only must we think globally, we must also act globally through culturally sensitive project management. It is impossible for any organization to run any large project nowadays without some aspect of international involvement. Industry 4.0 depends on good project management across organizational and global boundaries. Project management is the pursuit of organizational goals within the constraints of time, cost, and performance expectations. Project expectations can be in terms of physical products, service, and results. Global interactions of business and industry imply that global situational awareness be instituted for international project management. Situational awareness is the process of recognizing and appreciating unique factors that define the operating characteristics in given geographical location. These factors can range widely from one country to another. What are believed to be normal operating conditions in home country operations may be taboo, illegal, restricted, or forbidden in another country. Far too often, organizations don't pay enough attention to this fact. Training programs, briefings, seminars, and sensitivity role plays are often used to prepare personnel for overseas assignments. But the fact remains that global awareness failure points still exist among organizations operating overseas, away from home base – whether from the East to the West or from the West to the East or elsewhere.

We can just read the world news headlines on any given day and we will find global project interconnections in oil, autos, industrial credits, construction, and so on. Such mega deals are not limited to the G8, G10, G12, or G20. Even the little non-G economies are in the mix of international projects. What connects all of these is global project management. So pervasive has the need for project management become that the Project Management Institute (PMI) proclaims that "Worldwide, organizations will embrace, *value, and utilize* project management *and attribute* their success to it." Indeed, project management is the common language that all organizations can embrace particularly when operating at the global level.

DIVERSITY OF CULTURES IN INDUSTRY 4.0

The increased interface of cultures through international project outsourcing is gradually leading to the emergence of hybrid cultures in many developing countries. A hybrid culture derives its influences from diverse factors, where there are differences in how the local population views education, professional loyalty, social alliances, leisure pursuits, and information management. A hybrid culture poses a big challenge to managing internationally-outsourced projects. But the problem can be mitigated by a strategic program of situational awareness. Presented below are some pros and cons of cultural enmeshing at the global level with respect to Industry 4.0.

Pros:

- Global awareness for competition
- Culturally diverse workforce
- Access of international labor force
- Intercultural peace and harmony
- Facilitation of social interaction
- Closing of trade gaps

Cons:

- Subjugation of one culture to the other
- Cultural differences in work ethics
- Loss of cultural identity
- Bias and suspicion
- Confusion about cultural boundaries
- Non-uniform business vocabulary

MARKET COMPETITION UNDER INDUSTRY 4.0

Many western industries are unable to compete globally on the basis of labor cost, which is where the pursuit of competitiveness is often directed. The competitive advantage for many manufacturers will come from appropriate infusion of technology into the enterprise. Strategic research, development, and implementation of technological innovations will give manufacturers the edge needed to successfully compete globally. In spite of the many decades of lamenting about the future of manufacturing, very little has been accomplished in terms of global competitiveness. Part of the problem is the absence of a unified project management approach. Managing global and distributed production teams requires a fundamental project systems approach.

One recommendation here is to pursue more integrative linkages of technical issues of production and the operational platforms available in industry. Many concepts have been advanced on how to bridge the existing gaps. But what is missing appears to be a pragmatic project-oriented roadmap that will create a unified goal that adequately, mutually, and concurrently, addresses the profit-oriented focus of practitioners in industry and the knowledge-oriented pursuits of researchers in academia. The problems embody both scientific and management issues. Many researchers have not spent sufficient time in industry to fully appreciate the operational constraints of industry. Hence, there is often a disconnection between what research dictates and what industry practice requires. An essential need is the development of a global project roadmap. One aspect that is frequently ignored in this respect is the set of human-cultural factors operating in globally distributed work teams. A culturally-sensitive project management approach will enable an appreciation of this crucial component of global projects.

In order to enhance global awareness, the preparation approach that can be more effective and sustainable is to address specific sets of global-awareness questions without making any prior assumptions in each and every overseas assignment case. Even previous international experience can become out of phase become local situations can change of time. The questions categorized below can be helpful.

Pre-assignment Planning for Industry 4.0

- What is the organizational mission in the overseas location?
- Is the personnel aware of the global situation as it relates to the country or region of operation?
- Will the assignment be as a team, individual operation, or organizational attachment?
- What legal status will be associated with the project during the overseas assignment?
- What country locations are involved in the assignment?
- What healthcare situations and medical services are available during the assignment?
- What personal family arrangements and precautions are relevant for the overseas assignment?

- What prudent precautions are possible? Will update? Health insurance? Medical records?
- What is the currency exchange rate? Location and access to bureau de change?

Sample questions for local conditions

- What kind of local environment will be in effect at the assignment location? Urban city? Rural? Jungle? Mountain? Desert?
- What is the typical weather pattern? Seasons? Cold? Hot? When, where and how long?
- What are the general living conditions? Access to clean water? Bathroom facilities?
- What are the domestic transportation options? Cost? Reliability? Access? Rivers? Trails? Trains? Shuttle buses? Commuter planes?
- What is the level of environmental consciousness?
- What is the local social hierarchy?

Sample questions for cultural nuances

- Is there a compilation of acceptable and unacceptable behavior in the assignment location?
- What are the political beliefs?
- What are the typical customs?
- What are the prevailing religious beliefs and restrictions?
- Are there guidelines for use of body language and gestures?
- What are the offensive symbols and words?

Sample questions for geographic location

- What are the bordering regions, areas, and/or countries to the assignment location?
- What is the operating terrain? Flat? Hilly? Road system?
- Is there access to rivers, oceans, beaches, etc.?
- What is the vegetation like?
- What are the major industries? Manufacturing factories? Service industry?
- Are there farming establishments? Access to fresh foods?

Sample general questions

- What is the social environment?
- Is there night life? Is it accessible? It is safe?
- What is the crime statistics?

- What mode of law enforcement?
- What is the restriction on currency importation and exportation?
- What is the local manpower level? Skills? Availability? Cost? Dependability?
- What is the educational opportunity? Access? Affordability? Level?
- What international support organizations are available? Red Cross?
- What is the human rights record of the locality?

If questions such as the above are addressed forthrightly, international projects can run more effectively, more efficiently, and more successfully. Typical sources of barriers to a comprehensive global awareness include the following:

Denial: This presents the danger of not fully recognizing the problem of global awareness. A person may dismiss the problem or minimize the gravity of the problem.

Past experience: This has the danger of luring the personnel into an erroneous decision based on a past or similar experience.

Complacency: This has the danger of feeling of overconfidence about knowing what obtains in the assignment location, thereby dismissing the necessity for proactive preparation.

Insufficient information: This has the danger of misinterpretation due to limited or unreliable information about the assignment location.

GLOBAL LABOR COSTS IN INDUSTRY 4.0

Some cultures, by their inherent nature, offer lower labor costs for international operations. Consequently, the search for a competitive operating site might take an organization to a culture that is totally different from the base station. The pervasiveness of fast routing of information makes it easy for a business to find a seemingly receptive overseas site to relocate operations. The consequence of such relocation is a transfer of culture in either direction, often with limited situational awareness. Many organizations don't fully appreciate the differences in the operating cultures. They pay attention only to the physical and economic aspects of their operations. But more often than not, the cultural shock and unsuccessful assimilation (from either end of the culture transfer) can lead to project failures. Many of the economically underserved countries in Asia, Africa, Oceania, and Latin America are frequent targets to international project development. Those countries have common characteristics, such as highly dependent economies (devoted to producing primary products for the developed world), traditional, rural social structures, high population growth, and widespread poverty. Certain characteristics exist that may constitute barriers to successful global projects. Some of these are:

- Limited access to information (substandard telecommunications infrastructure)
- Politically-induced trade barriers
- Cultural norms that impede free flow of information
- Existence and abundance supply of cheap, albeit untrained, workforce
- Orientation of manpower toward artisan and apprenticeship labor

GLOBAL WORKFORCE CONSTRAINTS

Most project outsourcing points are located in developing and underdeveloped nations. These locations often have repressive cultures that are replete with norms that the western world would find unacceptable. A cultural bridge usually is missing between the developed nations and the developing nations with respect to workforce capabilities. Thus, there are increasing cultural and economic disparities between global business partners. Some of the local issues to be factored into global situational awareness programs include:

- Poverty
- Pollution
- Disease
- Inferior health services
- Political oppression
- Gender biases
- Economic and financial scams
- Wealth Inequities
- Social permissiveness among the elite

Project personnel posted to these regions are shocked by the level of cultural differences that they experience. In some cases, they maintain a "laissez faire" and hands-off attitude. But there have also been cases where some of the personnel take advantage of the loose culturally acceptable social contacts that could be to the detriment of an international project. Of particular consideration is what the COVID-19 pandemic has done to workforce profiles across the board. In the face of the pandemic, organizations responded by instituting virtual operations, remote work, and telework. Unfortunately, after dwindled, workers are reluctant to return to in-person work within the confines of the typical workplace in the employer's premises. This has a continuing impact on the supply chain and value chain for Industry 4.0.

CULTURE-BASED HIERARCHY OF NEEDS

The psychology theory of "Hierarchy of Needs" proposed by Abraham Maslow in his 1943 paper, "A Theory of Human Motivation," still governs how different cultures respond along the dimensions of global project expectations. A culturally-induced disparity in the hierarchy of needs implies that we may not be able to fulfill our project responsibilities along the global spectrum of international projects. In a culturally different workforce, the specific levels and structure of the needs may be drastically different from the typical mode observed in western culture. Maslow's hierarchy of needs consists of five stages:

1. **Physiological Needs**: These are the needs for the basic necessities of life, such as food, water, housing, and clothing (i.e., Survival Needs). This is the level where access to money is most critical.
2. **Safety Needs**: These are the needs for security, stability, and freedom from physical harm (i.e., Desire for a safe environment).
3. **Social Needs**: These are the needs for social approval, friends, love, affection, and association (i.e., Desire to belong). For example, social belonging may bring about better economic outlook that may enable each individual to be in a better position to meet his or her social needs.
4. **Esteem Needs**: These are the needs for accomplishment, respect, recognition, attention, and appreciation (i.e., Desire to be known).
5. **Self-Actualization Needs**: These are the needs for self-fulfillment and self-improvement (i.e., Desire to arrive). This represents the stage of opportunity to grow professionally and be in a position to selflessly help others.

In an economically-underserved culture, most workers will be at the basic level of physiological needs; and there may be cultural constraints on moving from one level to the next higher level. This fact has an implication on how cultural interfaces may fail between host and guest nations involved in a global project.

INDUSTRY 4.0 PROJECT SUSTAINABILITY

Sustainability – it isn't just for the environment. Project sustainability on the global front is as much a need as the traditional components of project management spanning planning organizing, scheduling, and control. For international projects, preemption of cultural conflicts is far better than post-occurrence remedies. Proactive pursuit of project best practices can pave the way for project success on a global scale.

Cultural infeasibility is one of the major impediments to project outsourcing in an emerging economy. The business climate of today is very volatile. This volatility, coupled with cultural limitations, creates problematic operational elements for global projects located in a developing country. For example, an inquiry or revelation of personal information is viewed as taboo in many developing countries. Consequently, this may impede the collection, storage, and distribution of workforce information that may be vital to the success of global project management. Global situational awareness and its best practices are essential for any organization engaged in international projects. In these days of globalization, nothing happens without international cooperation; and cooperation cannot succeed without effective situational awareness on each side of the cooperating partners.

EXTRAPRENEURSHIP CONCEPT
FOR INDUSTRY 4.0

"United we manufacture" is a cliché that aptly describes how to enhance manufacturing in a tight economy. The concept of "Extrapreneurship" is a collaborative approach to manufacturing partnership in a tight economy that is governed by rapidly-changing technologies. Survival of all as a group means survival of each and every entity in the group. We need a healthy manufacturing base in order to have an overall viable economy. Even the booming service enterprise industry depends on outputs from the manufacturing sector. There is an increasing demand for all things manufactured – from milling, turning, drilling, grinding, cutting, forging, plastic injection molding, finishing, and casting to welding and more; we all depend on mutual manufacturing outputs. Manufacturing lays the foundation for economic advancement. It is, thus, imperative to jointly develop strategies for enhancing that unifying sector of the global economy. Without creative solutions, the future of independent manufacturing in a tight economy is very bleak. What ails one manufacturer often affects other manufacturers either directly or indirectly. Manufacturers are all enmeshed in the global system of market movements. Cooperating partnership among those similarly affected is one way for all to survive. The recent coordinated expedition of the big three of the US auto industry in search of government bailout is illustrative of the need for competitors to partner even in the face of fierce market rivalry. If General Motors can boost the chances for the survival of Chrysler, it is directly boosting the industry upon which its own survival rests. This article presents extrapreneurship as a partnering strategy for manufacturers in a tight economy.

WHAT IS EXTRAPRENEURSHIP?

Terms such as "coopetition" have been used in the past to describe strategic alliances between competitors. Left to its own devices, cooperating competition can falter and degrade over time, particularly when market push comes to market shove. In this article, we take a step further to introduce *extrapreneurship* as an extension to entrepreneurship, intrapreneurship, and other "preneurships" currently embraced by business and industry. Entrepreneurship sets up independent business pursuits that compete in the open market. Intrapreneurship sets up an internal semi-autonomous business unit of an organization that may operate under a different business model from the parent company. Extrapreneurship sets up a physical presence of one manufacturer within the operation of another cooperating manufacturer. Manufacturers must understand and embrace the interplay of organizational structures of entrepreneurship, intrapreneurship, and extrapreneurship, with respect to Industry 4.0. It should be emphasized that extrapreneurship is not conventional outsourcing. It is a direct set up of operation within the facility of the cooperating manufacturer. It allows a manufacturer to have a presence (or even a product) in an area for which it does not have internal core manufacturing competence.

Although elements of what extrapreneurship entails are being practiced by many organizations (e.g., customer site co-location), there has not been a formal or unified formulation as a manufacturing initiative. Extrapreneurship sets up a unit of one competing-but-cooperating company within the premises of another company. It is the next best thing to direct acquisition or merger. This approach indicates that the collective success of the manufacturing sector requires a systems-oriented synergistic approach. Not all manufacturers are fully versed in all segments of their operations. Thus, embracing (or even housing) a unit of a competitor that brings added value is a strategy that creates mutual manufacturing support.

KEY COMPONENTS OF EXTRAPRENEURSHIP

There are several concomitant components of extrapreneurship; however, two key components are discussed here:

1) Strategic Partnering
2) Cooperation

Unlike conventional strategic business partnering, extrapreneurial strategic partnering has a requirement for physical co-location of units of the partnering organizations. Strategic alliance is defined as a formal alliance or "joining of forces" between two or more independent organizations for the purpose of meeting mutual business goals. Each partner in the alliance has something to bring to the "table," such as products, supply chain, distribution network, manufacturing capability, funding, capital equipment, operational expertise, know-how, or intellectual property. Strategic partnering represents cooperation whereby synergy ensures that each partner derives benefits beyond normal independent operation.

The co-location strategy of extrapreneurship is what facilitates coordination and implementation of the ideals of cooperating. It represents partnership outreach of business-to-business collaboration that borders on subsidiary relationship. Manufacturing organizations must leverage existing structures of entrepreneurship and intrapreneurship to create sustainable extrapreneurship relationships. This triangular strategy conveys the goal of achieving communication, cooperation, and coordination of manufacturing initiatives. Communication facilitates cooperation while cooperation makes operational coordination possible. With a strategic project management approach, we can build constructive extrapreneurial relationships rather than adversarial relationships among manufacturers.

STRATEGIC PARTNERING FOR EXTRAPRENEURSHIP

While there are pros and cons of partnering, the advantages often outweigh the downsides. Advantages of strategic partnering include the following:

1) It allows each partner to concentrate on operations that best match its capabilities.
2) It permits partners to learn from one another and develop competencies that may be readily utilized elsewhere.
3) It facilitates synergy that increases the outputs of partners' resources and competencies.

The key characteristic requirement for extrapreneurship is the physical co-location of units of the partnering organizations. Strategic partnering can be executed in a step-wise formation as described below.

Step 1: Strategy Development. This requires a feasibility study of the proposed alliance with respect to objectives, rationale, people, technology, process, resource base, management, challenges, and conflict resolution strategies. Strategy development requires aligning the objectives of the alliance with the overall corporate vision and mission. A key part of strategy development is an internal SWOT analysis to document strengths, weaknesses, opportunities, and threats.

Step 2: Partner Assessment. This involves assessing a potential partner's capabilities, performing benchmarking analysis, developing tactics to accommodate a partner's management styles, developing criteria for partner selection, and understanding mutual resource requirements.

Step 3: Agreement Negotiation. This involves determining prospective partners' relative objectives, contributions, rewards, information protection, team composition, performance assessment procedures, policies, procedures, and termination clauses.

Step 4: Strategy Implementation. This involves actual operation of the alliance between the partners. It requires management's commitment to resource allocation, linking of budgets, coordination of priorities, schedule development, tracking, control, and reporting. This is an area where project management techniques could be most useful. Many alliances written so beautifully on paper often falter due to lax implementation strategies. Partnering strategy implementation is of four different types:

a) **Joint venture**: This is a strategic alliance in which two or more partners create a legally independent company to share some of their resources and capabilities to develop a mutual competitive advantage in the market place.

b) **Equity strategic partnership**: This is an alliance in which two or more partners own different percentages of the company they have formed by combining some of their resources and capabilities to create a mutual competitive advantage in the market place.

c) **Non-equity strategic partnership**: This is an alliance in which two or more partners develop a contractual relationship to share some of their unique resources and capabilities to create mutual competitive advantage.

d) **Global strategic partnership**: This is a working partnership between companies (often more than 2) across national boundaries and/or across industries. This is often formed between a company and a foreign government.

Step 5: Termination. All good things eventually come to an end. Partners must be realistic about this fact-based cliché. While sustainability is of utmost importance in alliance formation, what needs to be terminated must be terminated when the time comes. Force-feeding an alliance just for the sake of sustaining a relationship could only lead to a waste of time, squandering of resources, and lost opportunities. It is time to terminate and alliance when its objectives have been met or cannot be met, or when a partner adjusts priorities or divert resources toward other initiatives.

COOPERATION FOR EXTRAPRENEURSHIP

Cooperation is a basic requirement for resource interaction and integration between partners. More projects fail due to a lack of cooperation and commitment than any other project factors. To secure and retain the cooperation of extrapreneurial partners, the most positive aspects of a proposed partnership should be the first items of extrapreneurship communication. Such structural communication can pave the way for acceptance of the proposal and subsequent cooperation. For extrapreneurial partnering, there are different types of cooperation that can be in effect, as summarized below:

- **Functional Cooperation**: This is cooperation induced by the nature of the functional relationship between two partners. The two partners may be required to perform related functions that can only be accomplished through mutual cooperation.
- **Socially-responsible Cooperation**: This is the type of cooperation effected by a socially-responsible relationship between two partners. This is particularly common for activities that may impact the environment. The socially-responsible relationship motivates cooperation that may be useful in executing extrapreneurial partnership.
- **Regulatory Cooperation**: This is usually cooperation that is based on regulatory requirements. It is often imposed through some legal authority and expectations. In this case, the participants may have no choice other than to cooperate.
- **Industry Cooperation**: This is cooperation that is fueled by the need to comply with industry standards and build a consensus to advance the overall industry in which partners find themselves.
- **Market Cooperation**: In order for each player in the market to thrive, the overall market must be vibrant. For this reason, market cooperation involves partnering of market players to advance market health.
- **Administrative Cooperation**: This is cooperation brought on by administrative requirements that make it imperative that two partners work together toward a common goal, such as market survival.
- **Associative Cooperation**: This is a type of cooperation that may be induced by collegiality. The level of cooperation is determined by the prevailing association that exists between the partners. Industry associations often cooperate under this approach.
- **Proximity Cooperation**: This type of cooperation may be viewed as "Silicon-Valley orientation" whereby organizations located within the same geographical setting form cooperating alliances to pursue mutual market interests. Being close geographically makes it imperative that the partners work together.
- **Dependency Cooperation**: This is cooperation caused by the fact that one partner depends on another partner for some important aspect of its operation and business survival. Such dependency is usually of a mutual two-way structure. One partner depends on the other partner for one thing while the latter partner depends on the former partner for some other thing.

- **Imposed Cooperation**: In this type of cooperation, external forces are used to induce cooperation between partners. This is often the case with legally-binding requirements.
- **Natural Cooperation**: This is applicable for cases where the two partners have no way out of cooperating. Physical survival requirements often dictate this type of cooperation.
- **Lateral Cooperation**: Lateral cooperation involves cooperation with peers and immediate contemporaries in the market place. Lateral cooperation is often possible because existing lateral relationships create an environment that is conducive for mutual exchanges of information and operational practices. An example is the recent bailout pursuit by the big three of the US auto industry.
- **Vertical Cooperation**: Vertical or hierarchical cooperation refers to cooperation that is implied by the hierarchical structure of the market in which the partners operate. For example, subsidiaries are expected to cooperate with their vertical parent organizations.

Whichever type of cooperation is available in an extrapreneurial environment, the cooperative forces should be channeled toward achieving mutual goals. A documentation of prevailing level of cooperation is useful for winning further support and sustaining joint pursuits. Clarification of organizational priorities will facilitate personnel cooperation. Relative priorities of multiple partners should be specified so that a venture that is of high priority to one segment of the extrapreneurial engagement is also of high priority to all partners within the endeavor.

Competition can, indeed, be used as a mechanism for extrapreneurial cooperation. This happens in an environment where constructive competition paves the way for cooperation. This is sometimes designated as "coopetition" whereby the pseudo-competing organizations form a partnership to advance their mutual interests. To ensure success, cooperation must be SMART (Specific, Measurable, Aligned, Realistic, and Timed). Some guidelines for securing extrapreneurial cooperation are presented below:

- Establish achievable goals for the extrapreneurial initiative
- Clearly outline individual commitments that are required
- Integrate extrapreneurship priorities with existing organizational priorities
- Anticipate and mitigate potential sources of conflict
- Alleviate skepticism by documenting the merits of the extrapreneurial initiative

MANAGING EXTRAPRENEURIAL PROJECTS

Project management is required for getting things done and moving concepts to reality. This author defines project management as the process of managing, allocating, and timing resources to achieve a given goal in an efficient and expeditious manner. The best of concepts is nothing if it cannot be practiced. The concept of extrapreneurship uses project management strategies as the basis for practical accomplishment of what cooperation sets out to achieve. Thus, it goes beyond mere buzzword. Recognizing the category of products expected from an enterprise is an essential part of managing extrapreneurial alliances more effectively. Project outputs are categorized into the classes below. This makes the application of project management process applicable to every undertaking because each effort is expected to generate an output in one or more of the following output categories.

- Product: Physical products, e.g., new hospital facility
- Service: Business process, e.g., new operating procedure
- Result: Knowledge creation, e.g., research result, education, training output

In evaluating the potential for extrapreneurship in manufacturing, partners must evaluate their respective categories of expected outputs and how to manage the outputs. Formal application of project management also creates an effective extrapreneurial audit process that enhances overall business operations through the following:

- Customer focus from an external viewpoint
- Industry leadership benchmarking and development
- Involvement of people across organizational boundaries
- Diversification of business processes
- Systems view of the market
- Pathway to continual improvement along external market trends
- Mutual dependency leading to win–win commitment

Axioms of extrapreneurship for manufacturing operations include the following:

1. Manufacturers must communicate for mutual exchange of information that affects their operations.

2. Manufacturers must cooperate for the purpose of synergy that promotes win-win results for all.
3. Manufacturers must coordinate to leverage their respective strengths for a unified advancement of the market.

BENEFITS OF EXTRAPRENEURSHIP FOR INDUSTRY 4.0

As a business technique, extrapreneurship opens the door to many benefits. Examples include:

- Promotes mutual industry understanding
- Presents a unified view of the market
- Exchanges industry's best practices
- Enhances true collaboration
- Fosters proximity-based cooperation
- Shortens coordination points
- Facilitates real-time and on-site communication
- Encourages operational transparency
- Advocates industry teamwork
- Builds non-adversarial relationships
- Removes operational waste (lean-centric)
- Creates avenues for linking internal and external checks and balances
- Provides effectiveness of controls in the organization and its extrapreneurial units.

The benefits of extrapreneurship are not limited to the business or manufacturing environment. Variants of the approach can benefit other enterprises such as university-to-university alliances and retail-to-retail exchanges. At the global level, extrapreneurship strategies can be used to extend operations to other parts of the world. This will facilitate transnational movement of goods and services between nations, beyond national boundaries. It can also fuel international business advancement between collaborating nations in a way that can promote mutual economic dependency that is essential for achieving peace. Cost, schedule, and quality requirements are key reasons for organizations to cooperate and explore potentials for extrapreneurship.

CONCLUSION

Beyond the technological backbone of Industry 4.0, there are many managerial aspects that must be recognized. If all manufacturing entities thrive under the framework of Industry 4.0, it means each manufacturing unit will prosper. To kill off a competitor does not necessarily mean that the killer survives in the industry. If the industry fails, each and every entity in the industry would be doomed. Cases in point include the US fabrics and steel industries, which suffered with massive offshoring exodus in the 1990s. Many other manufacturing segments are at risk of annihilation if innovative business survival strategies are not developed, embraced, and pragmatically implemented. The ripple effect of the depressed world economy is spreading throughout every corner of the manufacturing sector, but Industry 4.0, implemented from a systems perspective, can come to the rescue. It behooves all manufacturers to embrace elements of Industry 4.0 and be prepared for future developments in order to ensure the survival of "system of the whole." The next chapter presents the fundamentals of systems-based operations.

REFERENCES

Badiru, Adedeji B. (2009), "Twin Fates: Partnerships will keep manufacturers' doors open," *Industrial Engineer, March*, pp. 40–44.

Badiru, Adedeji B. editor (2014), *Handbook of Industrial & Systems Engineering*, 2nd Edition, Taylor & Francis/CRC Press, Boca Raton, FL.

Badiru, Adedeji B. (2019), *Project Management: Systems, Principles, and Applications*, 2nd Edition, Taylor & Francis/CRC Press, Boca Raton, FL.

Badiru, Adedeji B. (2020), *Innovation: A Systems Approach*, Taylor & Francis/CRC Press, Boca Raton, FL.

Badiru, Adedeji B. (2022), "Quality Insight: Product Quality Assurance Under Industry 4.0," *International Journal of Quality Engineering and Technology*, DOI: 10.1504/IJQET.2023.10048108

Badiru, Adedeji B. (2023), *Systems Engineering Using DEJI Systems Model: Design, Evaluation, Justification, and Integration with Case Studies and Applications*, Taylor & Francis/CRC Press, Boca Raton, FL.

Badiru, Adedeji B., Oye Ibidapo-Obe, and Babs J. Ayeni (2019), *Manufacturing and Enterprise: An Integrated Systems Approach*, Taylor & Francis/CRC Press, Boca Raton, FL.

Cretu, Ana (2020), "Industry 4.0 Essentials: Storyline, Technologies, Adoption," *Industrial Automation, Industry 4.0*, https://www.qwertee.io/blog/industry-40-essentials-storyline-technologies-adoption/, Accessed 20 January 2023.

Elangovan, Uthayan (2022), *Industry 5.0: The Future of the Industrial Economy*, CRC Press, Taylor & Francis, Boca Raton, FL.

Foidl, Harald and Michael Felderer (2016), *Research Challenges of Industry 4.0 for Quality Management, Conference Paper*: DOI: 10.1007/978-3-319-32799-0_10

Zaidin, Nur Hanifa Mohd, Muhammad Nurazri Diah, Po Hui Yee, and Shahryar Sorooshian (2018), "Quality Management in Industry 4.0 Era," *Journal of Management and Science 8*(2): 82–91. DOI: 10.26524/jms.2018.17

Elements of Systems 4.0

2

INTRODUCTION TO SYSTEMS APPROACH

A system is represented as consisting of multiple parts, all working together for a common purpose or goal. Systems can be small or large, simple or complex. Small devices can also be considered systems. Systems have inputs, processes, and outputs. Systems are usually explained using a model for a visual clarification inputs, process, and outputs. A model helps to illustrate the major elements and their inter-relationships. Figure 2.1 illustrates the basic model structure of a system, with a specific reference to the requirements of Industry 4.0. For the purpose of this book, we refer to this interplay as Systems 4.0. Systems 4.0 is the application of the general principles of systems methodology to the requirements of Industry 4.0.

Systems engineering is the application of engineering tools and techniques to the solutions of multifaceted problems through a systematic collection and integration of parts of the problem with respect to the life cycle of the problem. It is the branch of engineering concerned with the development, implementation, and use of large or complex systems. It focuses on specific goals of a system considering the specifications, prevailing constraints, expected services, possible behaviors, and structure of the system. It also involves a consideration of the activities required to ensure that the system's performance matches specified goals. Systems engineering addresses the integration of tools, people, and processes required to achieve a cost-effective and timely operation of the system. Among the features of this book are a solution to multifaceted problems, a holistic view of a problem domain, applications to both small and large problems, the decomposition of complex problems into smaller manageable

DOI: 10.1201/9781003312277-2

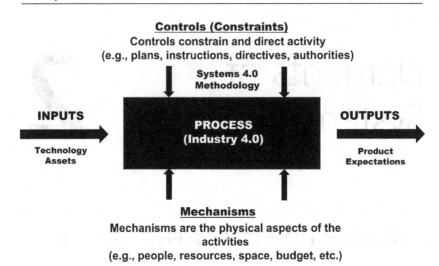

FIGURE 2.1   Systems 4.0 Framework for Industry 4.0.

chunks, direct considerations for the pertinent constraints that exist in the problem domain, systematic linking of inputs to goals and outputs, explicit treatment of the integration of tools, people, and processes, and a compilation of existing systems engineering models. A typical decision support model is a representation of a system, which can be used to answer questions about the system. While systems engineering models facilitate decisions, they are not typically the conventional decision support systems.

The end result of using a systems engineering approach is to integrate a solution into the normal organizational process. For that reason, the DEJI systems model is desired for its structured framework of Design, Evaluation, Justification, and Integration.

SYSTEMS ENGINEERING MODELS

Several systems engineering models are available in practice (Satzinger et al., 2016; Colombi and Cobb, 2009; Furness, 2018; Haas et al., 2009; Henning and Walter, 2012: Jia-Ching, 2011; London, 2012; McMurtry, 2013; DAU, 2017; Tutorials Point, 2018). In many cases, the applications are customized for internal organizational applications and are not fully documented in the open literature. The most common models include the Waterfall model, the

V-model, the Spiral model, the Walking Skeleton model, and others. Many of them originated in the software development industry. Selected ones are described below (Badiru, 2019).

THE WATERFALL MODEL

The Waterfall model, also known as the linear-sequential life-cycle model, breaks down the Systems Engineering (SE) development process into linear sequential phases that do not overlap one another. The model can be viewed as a flow-down approach to engineering development. The Waterfall model assumes that each preceding phase must be completed before the next phase can be initiated. Additionally, each phase is reviewed at the end of its cycle to determine whether or not the project aligns with the project specifications, needs, and requirements. Although the orderly progression of tasks simplifies the development process, the Waterfall model is unable to handle incomplete tasks or changes made later in the life cycle without incurring high costs. This makes sense for the waterfall model since water normally flows downward, unless forced to go upward through a pumping device, which could be an additional cost. Therefore, this model lends itself better to simple projects that are well defined and understood.

Case Study: This model was instrumental to updating the Conway Regional Medical Center in Arkansas. In 2010, the regional center still did not have an electronic database and information management system for its home health care patients. Consequently, the hospital applied the Waterfall method to their acquisition of software to handle their needs. First, the hospital management defined their problem as needing a way to maintain a database of documentation and records for the home health care patients. The hospital then elected to buy, rather than design, the software necessary for this project. After defining their system requirements, the hospital's administration team purchased what they evaluated to be the most suitable software option. However, the hospital performed systems testing before integrating the software into the home healthcare system. Upon completion of testing, the hospital found that the code needed to be updated once every six weeks. This update was factored into their operation and maintenance plan for use of this new system. The system was finally deemed a resounding success once the system was implemented, tested, and the operations and maintenance schedules were created. Through the use of the waterfall model, the software got on track for installation as the primary home health care software for the Conway Regional Medical Center.

THE V-MODEL

The V-model, or the verification and validation model, is an enhanced version of the Waterfall model that illustrates the various stages of the system life cycle. The V-model is similar to the Waterfall model in that they are both linear models whereby each phase is verified before moving on to the next phase. Beginning from the left side, the V-model depicts the development actions that flow from concept of operations to the integration and verification activities on the right side of the diagram. With this model, each phase of the life cycle has a corresponding test plan that helps identify errors early in the life cycle, minimize future issues, and verify adherence to project specifications. Thus, the V-model lends itself well to proactive defect testing and tracking. However, one drawback of the V-model is that it is rigid and offers little flexibility to adjust the scope of a project. Not only is it difficult, but it is also expensive to reiterate phases within the model. Therefore, the V-model works best for smaller tasks where the project length, scope, and specifications are well defined.

Case Study: The V-model was used with great success to establish the Chattanooga Area Regional Transportation Authority (CARTA), which was one of the first Smart Transport systems in the United States. The V-model was used to guide the design of the new system and to integrate this into the existing system of buses, electric transport, and light rail cars. The new smart system introduced a litany of features such as customer data management, automated route scheduling to meet demand, automated ticket vending, automated diagnostic maintenance system, and computer-aided dispatch and tracking. These features were revolutionary for a mid-sized metropolitan area. CARTA was able to maintain their legacy transport, while integrating their new system. They were able to do this by splitting the V-model into separate sections. CARTA had a dedicated team to manage the flat portions of the V. This portion encapsulated the legacy transportation systems. CARTA then had individual teams focus on the definition, test, and integration of all the new components of the system. Separating the two sides – legacy and innovation – enabled CARTA to maintain valuable functionalities while adding new features to their system that enhanced usability, safety, and efficiency.

SPIRAL MODEL

The Spiral model is similar to the V-model in that it references many of the same phases as part of its color coordinated slices, which indicates the project's stage of development. This model enables multiple flows through the cycle to build a better understanding of the design requirements and engineering complications. The reiterative design lends itself well to supporting popular model-based systems engineering techniques such as rapid prototyping and quick failure methods. Additionally, the model assumes that each iteration of the spiral will produce new information that will encourage technology maturation, evaluate the project's financial situation, and justify continuity. Consequently, the lessons learned from this method provides a data point with which to improve the product. Generally, the Spiral model meshes well with the defense life cycle management vision and integrates all facets of design, production, and integration.

The Spiral model is the foundation of the RQ-4 Global Hawk Operational Management and Usage platform. The Global Hawk was phased into operation in six distinct spirals, with each spiral adding new capabilities to the airframe. The first spiral was getting the aircraft in the sky and having a support network to keep it in the air. Everything from pilots to maintainers was optimized to keep the Global Hawk in the air as much as possible. The subsequent phases added imagery (IMINT), signal (SIGINT), radar, and survivability capabilities to the airframe. Each of these capabilities was added one at a time in a spiral development cycle to ensure that each one was integrated into the airframe to the operational standard, and could be adjusted to meet this standard before moving on to the next capability. The benefit of incrementally adding capabilities in a spiral fashion greatly helped the Global Hawk stay on budget and schedule for operational rollout.

DAU MODEL

The new Defense Acquisitions University (DAU) model for Systems Engineering also originates from the V-Model. However, unlike the traditional V-model, it allows for process iteration similarly to the Spiral model. A unique attribute of the DAU process is that its life cycle does not need to be completed in order to gain the benefit of iteration. Whereas the Spiral model requires the life cycle process to be completed, the DAU model can refine and improve products at any point in its phase progression. This design is beneficial to making early-stage improvements, which helps Systems Engineers to avoid budgeting issues such as cost overruns. Moreover, the model allows for fluid transition between project definition (decomposition) and product completion (realization), which is useful in software production and integration. Overall, the DAU model is a fluid combination of the V-model and Spiral model.

This tailored V-model was used by the air force researchers to create a system to aid battlefield airmen in identifying friendly forces and calling in close air support with minimal risk to ground troops. The model was used to find an operational need and break it down into a hierarchy of objectives. The researchers used the hierarchy to design multiple prototypes that attempted to incorporate all of the stated objectives. They used rapid prototyping methods to produce these designs, and they were then tested operationally within battlefield airmen squadrons. Ultimately, the production of a Friendly Marking Device was achieved and this valuable capability was able to be delivered to the warfighter.

WALKING SKELETON MODEL

The Walking Skeleton model is a lean approach to incremental development, popularly used in software design. It centers on creating a skeleton framework for what the system is going to do and look like. This basic starting point of the system will have minimal functionality, and the systems engineer will work to add muscles to the skeleton. The first step creates a system that may do a very basic yet integral part of the final system design. For example, if one were to design a car using this method, the skeleton would be an engine attached to a chassis with wheels. Once the first basic step is done, the muscles begin to be added to the skeleton. These muscles are more refined and are added one at a time, meaning that each new feature of the system must be completed to add. Furthermore, it is highly recommended that the most difficult features of the system are the first muscles to be added. System components that take a lot of time, require contracting/outsourcing, or are the primary payload must be the first to be completed. This will become the heart of the skeleton and the rest of the architecture can be optimized to ensure the critical capability of the system is preserved and enhanced.

An example of engineering using the Walking Skeleton model are the Boston Dynamics walking dogs. The first thing that the engineers did for these robotic creatures was to create a power source and mobility framework. From there, the engineers were able to then go piece by piece and add more functionality to the project, such as the ability to open doors, pick up objects, and even carry heavy loads. The lessons they learned from adding muscles to their skeleton allowed them to move in leaps and bounds, and the benefits were felt across their entire network of products.

The Walking Skeleton technique varies with the system being developed. In the case of a client-server system it will be a single screen connected for navigating to database and back to screen.

In a front-end application system it acts as a connection between the platforms and compilation takes place for the simplest element of the language. In a transaction process it is walking through a single transaction.

The following are the techniques which can be used to create a Walking Skeleton:

- Methodology Shaping: Gathering information about prior experiences and using it to come up with the starter conventions. The following two steps are used in this technique:
  1. Project interviews
  2. Methodology shaping workshop
- Reflection Workshop: A particular workshop format for reflective improvement. In the reflection workshop team members discuss what is working fine, what improvements are required and what unique things will be added subsequently.
- Blitz Planning: Every member involved in project planning notes all the tasks on the cards, which will then be sorted, estimated, and strategized. Then the team decides on the resources such as cost and time and discusses what to do about the road blocks.
- Delphi Estimation: A way to come up with a starter estimate for the total project. A group consisting of experts is formed and opinions are gathered with an aim of arriving at highly accurate estimates.
- Daily Stand-ups: A quick and efficient way to pass information around the team on a daily basis. It is a short meeting to discuss status, progress, and setbacks. The aim is to keep meetings short. This meeting is to identify the progress and road blocks in the project.
- Agile Interaction Design: A fast version of usage-centered design, where multiple short deadlines are used to deliver working software without giving important considerations to activities of designing. To simplify the user–interface test LEET a record/Capture tool is used.
- Process Miniature: A learning technique as any new process is unfamiliar and time-consuming. When the process is complex more time is required for new team members to understand how different parts of the process fit. Time taken to understand the process is reduced with use of Process Miniature.
- Side-by-Side Programming: An alternative of pair programming is "Programming in pairs." Here two people work on one assignment by taking turns in providing input and mostly on a single workstation. It results in better productivity and cost consumed for fixing bugs is less. Programmers work without interfering in their individual assignments and review each other's work easily.
- Burn Charts: This tool is used to estimate actual and estimated amount of work against the time.

OBJECT ORIENTED ANALYSIS
AND DESIGN (OOAD)

Object Oriented Analysis and Design (OOAD) is an Agile Methodology approach towards systems engineering, and eschews traditional systems design processes. Traditional methods demand complete and accurate requirement specification before development; agile methods presume that change is unavoidable and should be embraced throughout the product development cycle. This is a foreign concept to many systems engineers that follow precise documentation habits, and would require an overhaul of project management architecture in order to work. If the necessary support is in place to allow for this approach, it works by grouping data, processes, and components into similar objects. These are the logical components of the system which interact with each other.

For example, customers, suppliers, contracts, and rental agreements would be grouped in a single object together. This object would then be managed by a single person with complete executive control over the data and relationships within. This approach is people-based, relying on the individual competencies and exquisite knowledge of their respective object. The systems manager then needs to link all of the people and their objects together to create the final system. The approach hinges upon each person perfecting their object that they are in charge of, and the systems engineer puts all of the pieces together. It puts all of the design control in the hands of the individual engineers. The most popular venue for use of this type of systems engineering is software engineering. It allows experts within their fields to focus on what they do best for a program. OOAD does not allow for convenient system oversight, process verification, or even schedule management, and as such makes it very difficult to get consistent project updates. While it may be conducive to small team projects, this method is unlikely to be feasible for any air force project of record.

COMPETENCY FRAMEWORK
FOR SYSTEMS ENGINEERING

The International Council on Systems Engineering (INCOSE) developed the INCOSE Systems Engineering Competency Framework (INCOSE SECF), which represents a worldview of five competency groupings with thirty-six competencies central to the profession of Systems Engineering. This includes evidence-based indicators of knowledge, skills, abilities and behaviors across five levels of proficiency. INCOSE SECF supports a wide variety of usage scenarios, including individual and organizational capability assessments. It enables organizations to tailor and derive their own competency models that address their unique challenges. The five competency groupings and their respective competencies are provided below:

- **Systems Engineering Management**
  1. Information Management
  2. Planning
  3. Monitoring and Control
  4. Risk and Opportunity Management
  5. Business & Enterprise Integration
  6. Decision Management
  7. Concurrent Engineering
  8. Configuration Management
  9. Acquisition and Supply
- **Professional**
  1. Communications
  2. Facilitation
  3. Ethics and Professionalism
  4. Coaching & Mentoring
  5. Technical Leadership
  6. Emotional Intelligence
  7. Negotiation
  8. Team Dynamics
- **Core Systems Engineering Principles**
  1. General Engineering
  2. Systems Thinking
  3. Capability Engineering
  4. Critical Thinking
  5. Systems Modeling and Analysis
  6. Lifecycles
- **Integrating**
  1. Quality
  2. Finance
  3. Project Management
  4. Logistics
- **Technical**
  1. Requirements Definition
  2. Systems Architecting

3. Transition
4. Operations and Support
5. Design (for specific needs)
6. Integration
7. Interfaces
8. Verification
9. Validation

Each competency in the SECF framework has five Proficiency Levels summarized below:

Awareness Proficiency: The person displays knowledge of key ideas associated with the competency area and understands key issues and their implications.

Supervised Practitioner Proficiency: The person displays an understanding of the competency area and has some limited experience.

Practitioner: The person displays both knowledge and practical experience of the competency area and can function without supervision on a day-to-day basis.

Lead Practitioner: The person displays extensive and substantial practical knowledge and experience of the competency area and provides guidance to others, including practitioners encountering unusual situations.

Expert: In addition to extensive and substantial practical experience and applied knowledge of the competency area, this individual contributes to and is recognized beyond the organizational or business boundary.

It should be noted that in many professional environments, it is believed that it takes about fifteen years of practical experience to become an expert in any particular professional pursuit.

SYSTEMS ATTRIBUTES, FACTORS, AND INDICATORS

In any systems approach, the systems analyst must be cognizant of the attributes, factors, and indicators that fully describe the overall system. For technical competency, the attributes, factors, and indicators of competency (knowledge and experience) are summarized below.

Competency Area: Technical (Requirements Definition)
Description: To analyze the stakeholder needs and expectations to establish the requirements for a system.

Purpose: The requirements of a system describe the problem to be solved (its purpose, how it performs, how it is to be used, maintained and disposed of and what the expectations of the stakeholders are).

For Awareness

- Describes different types of requirements (e.g. functional, non-functional, business etc.).
- Explains why there is a need for good-quality requirements.
- Identifies major stakeholders and their needs.
- Explains why managing requirements throughout the life cycle is important.

Supervised Practitioner

- Identifies all stakeholders and their sphere of influence.
- Assists with the elicitation of requirements from stakeholders.
- Describes the characteristics of good quality requirements and provides examples.
- Describes different mechanisms used to gather requirements.

Practitioner

- Defines governing requirements elicitation and management plans, processes, and appropriate tools and uses these to control and monitor requirements elicitation and management activities.
- Elicits and validates stakeholder requirements.
- Writes good-quality, consistent requirements.

Lead Practitioner

- Recognized, within the enterprise, as an authority in requirements elicitation and management techniques, contributing to best practice.
- Defines and documents enterprise-level policies, procedures, guidance, and best practice for requirements elicitation and management, including associated tools.
- Challenges appropriateness of requirements in a rational way.

Expert

- Recognized, beyond the enterprise boundary, as an authority in requirements elicitation and management techniques.
- Contributes to requirements elicitation and management best practice.
- Champions the introduction of novel techniques and ideas in requirements elicitation and management, producing measurable improvements

Competency Area: Systems Engineering Management (Risk and Opportunity Management)

Practitioner

- Defines governing risk and opportunity management plans, processes, and appropriate tools and uses these to control and monitor risk and opportunity management activities.
- Establishes a project risk and opportunity profile, including context, probability, consequences, thresholds, priority and risk action and status.
- Identifies, assesses, analyzes, and treats risks and opportunities for likelihood and consequence in order to determine magnitude and priority for treatment.
- Treats risks and opportunities effectively, considering alternative treatments and generating a plan of action when thresholds exceeds certain levels
- Guides supervised practitioners in Systems Engineering risk and opportunity management.

Lead Practitioner

- Recognized, within the enterprise, as an authority in Systems Engineering risk and opportunity management, contributing to best practice.
- Reviews and judges the tailoring of enterprise-level risk and opportunity management processes and associated work products to meet the needs of a project.
- Coordinates Systems Engineering risk and opportunity management across multiple diverse projects or across a complex system, with proven success.
- Establishes an enterprise risk profile, including context, probability, consequences, thresholds, priority and risk action and status.
- Coaches new and experienced practitioners in Systems Engineering risk and opportunity management.

For an effective application of SECF, all indicators must have the following properties:

- Start with action verbs.
- Are evidence-based.
- Show progressions from lower to higher levels of proficiency.
- Can be mapped to a combination of knowledge, skills, abilities, behaviors and experiences.
- Enable individuals to self-assess and acquire higher levels of proficiency.

SCOPE OF SYSTEMS ENGINEERING

The synergy between software engineering and systems engineering is made evident by the integration of the methods and processes developed by one discipline into the culture of the other. Researchers from software engineering (Boehm, 1994) and systems engineering (Rechtin, 1999 have extensively promoted the integration of both disciplines but have faced roadblocks that result from the fundamental difference between the two (Pandikow and Törne, 2001).

However, the development of systems engineering standards has helped the crystallization of the discipline as well as the development of COSYSMO. The first U.S. military standard focused on systems engineering provided by the first definition of the scope of engineering management (MIL-STD-499A, 1969). It was followed by another military standard that provided guidance on the process of writing system specifications for military systems (MIL-STD-490A 1985). These standards were influential in defining the scope of systems engineering in their time. Years later, the standard ANSI/EIA 632 *Processes for Engineering a System* (ANSI/EIA 1999) provided a typical systems engineering Work Breakdown Structure. The ANSI/EIA 632 standard was developed between 1994 and 1998 by a working group of industry associations, the International Council on Systems Engineering (INCOSE), and the U.S. Department of Defense with the intent to provide a standard for use by commercial enterprises, as well as government agencies and their contractors. It was designed to have a broader scope but less detail than previous systems engineering standards. Such lists provide, in much finer detail, the common activities that are likely to be performed by systems engineers in those organizations, but are generally not applicable outside of the companies in which they are created. In addition to organizational applicability, there are significant differences in different application domains, especially in space systems engineering (Valerdi et al., 2007).

The ANSI/EIA 632 standard provides the *what* of systems engineering through five fundamental processes: (1) Acquisition and Supply, (2) Technical Management, (3) System Design, (4) Product Realization, and (5) Technical Evaluation. These processes are the basis of the systems engineering effort profile developed for COSYSMO. The five fundamental processes are divided into 13 high-level process categories and further decomposed into 33 activities (see Badiru, 2019). This standard provides a generic list of activities which are generally applicable to individual companies, but each project should compare their own systems engineering Work Breakdown Structure (WBS) to the ones provided in the ANSI/EIA 632 standard to identify similarities and differences.

After defining the possible systems engineering activities used in COSYSMO, a definition of the system life cycle phases is needed to help bound the model and the estimates it produces. Because of the focus on systems engineering, COSYSMO employs some of the life cycle phases from ISO/IEC 15288 *Systems Engineering – System Life Cycle Processes* (ISO/IEC 2002). These phases were slightly modified to reflect the influence of the aforementioned model, ANSI/EIA 632. These life cycle phases help answer the *when* of systems engineering and COSYSMO. Understanding when systems engineering is performed relative to the system life cycle helps define anchor points for the model.

Life cycle models vary according to the nature, purpose, use and prevailing circumstances of the product under development. Despite an infinite variety in system life cycle models, there is an essential set of characteristic life cycle phases that exists for use in the systems engineering domain. For example, the *Conceptualize* phase focuses on identifying stakeholder needs, exploring different solution concepts, and proposing candidate solutions. The *Development* phase involves refining the system requirements, creating a solution description, and building a system. The *Operational Test & Evaluation* phase involves verifying/validating the system and performing the appropriate inspections before it is delivered to the user. The *Transition to Operation* phase involves the transition to utilization of the system to satisfy the users' needs via training or handoffs. These four life cycle phases are within the scope of COSYSMO. The final two were included in the data collection effort but did not yield enough data to perform a calibration. These phases are *Operate, Maintain, or Enhance*, which involves the actual operation and maintenance of the system required to sustain system capability, and *Replace or Dismantle*, which involves the retirement, storage, or disposal of the system.

SYSTEMS DEFINITIONS AND ATTRIBUTES

One definition of systems project management offered here is stated as follows:

> Systems project management is the process of using systems approach to manage, allocate, and time resources to achieve systems-wide goals in an efficient and expeditious manner.

The above definition calls for a systematic integration of technology, human resources, and work process design to achieve goals and objectives. There should be a balance in the synergistic integration of humans and technology. There should not be an over-reliance on technology, nor should there be an over-dependence on human processes. Similarly, there should not be too much emphasis on analytical models to the detriment of common-sense human-based decisions.

Systems engineering is growing in appeal as an avenue to achieve organizational goals and improve operational effectiveness and efficiency. Researchers and practitioners in business, industry, and government are all clamoring collaboratively for systems engineering implementations. So, what is systems engineering? Several definitions exist. Below is one quite comprehensive definition:

> **Systems engineering** is the application of engineering to solutions of a multi-faceted problem through a systematic collection and integration of parts of the problem with respect to the life cycle of the problem. It is the branch of engineering concerned with the development, implementation, and use of large or complex systems. It focuses on specific goals of a system considering the specifications, prevailing constraints, expected services, possible behaviors, and structure of the system. It also involves a consideration of the activities required to assure that the system's performance matches the stated goals. Systems engineering addresses the integration of tools, people, and processes required to achieve a cost-effective and timely operation of the system.

Logistics can be defined as the planning and implementation of a complex task, the planning and control of the flow of goods and materials through an organization or manufacturing process, or the planning and organization of the movement of personnel, equipment, and supplies. Complex projects represent a hierarchical system of operations. Thus, we can view a project system as a collection of interrelated projects all serving a common end goal. Consequently, we present the following universal definition:

> Project systems logistics is the planning, implementation, movement, scheduling, and control of people, equipment, goods, materials, and supplies across the interfacing boundaries of several related projects.

Conventional project management must be modified and expanded to address the unique logistics of project systems.

SYSTEMS CONSTRAINTS

Systems management is the pursuit of organizational goals within the constraints of time, cost, and quality expectations. The iron triangle model shows that project accomplishments are constrained by the boundaries of quality, time, and cost. In this case, quality represents the composite collection of project requirements. In a situation where precise optimization is not possible, there will have to be trade-offs between these three factors of success. The concept of iron triangle is that a rigid triangle of constraints encases the project. Everything must be accomplished within the boundaries of time, cost, and quality. If better quality is expected, a compromise along the axes of time and cost must be executed, thereby altering the shape of the triangle. The trade-off relationships are not linear and must be visualized in a multidimensional context. Scope requirements determine the project boundary and trade-offs must be done within that boundary. If we label the eight corners of the box as (a), (b), (c), ..., (h), we can iteratively assess the best operating point for the project. For example, we can address the following two operational questions:

1. From the point of view of the project sponsor, which corner is the most desired operating point in terms of combination of requirements, time, and cost?
2. From the point of view of the project executor, which corner is the most desired operating point in terms of combination of requirements, time, and cost?

Note that all the corners represent extreme operating points. We notice that point (e) is the do-nothing state, where there are no requirements, no time allocation, and no cost incurrence. This cannot be the desired operating state of any organization that seeks to remain productive. Point (a) represents an extreme case of meeting all requirements with no investment of time or cost allocation. This is an unrealistic extreme in any practical environment. It represents a case of getting something for nothing. Yet it is the most desired operating point for the project sponsor. By comparison, point (c) provides the maximum possible for requirements, cost, and time. In other words, the highest levels of requirements can be met if the maximum possible time is allowed and the highest possible budget is allocated. This is an unrealistic expectation in any resource-conscious organization. You cannot get everything you ask for to execute a project. Yet it is the most desired operating point for the project executor. Considering the two extreme points of (a) and (c), it is obvious that the project must be executed within some compromise region within the scope boundary. A graphical analysis can reveal a possible view of a compromise surface with peaks and valleys representing give-and-take trade-off points within the constrained box. The challenge is to come up with some analytical modeling technique to guide decision-making over the compromise region. If we could collect sets of data over several repetitions of identical projects, then we could model a decision surface that can guide future executions of similar projects. Such typical repetitions of an identical project are most readily apparent in construction projects, for example residential home development projects.

Systems influence philosophy suggests the realization that you control the internal environment while only influencing the external environment. The inside (controllable) environment is represented as a black box in the typical input–process–output relationship. The outside (uncontrollable) environment is bounded by a cloud representation. In the comprehensive systems structure, inputs come from the global environment, are moderated by the immediate outside environment, and are delivered to the inside environment. In an unstructured inside environment, functions occur as blobs. A "blobby" environment is characterized by intractable activities where everyone is busy, but without a cohesive structure of input-output relationships. In such a case, the following disadvantages may be present:

- Lack of traceability
- Lack of process control
- Higher operating cost
- Inefficient personnel interfaces
- Unrealized technology potentials

Organizations often inadvertently fall into the blobs structure because it is simple, low-cost, and less time-consuming; until a problem develops. A desired alternative is to model the project system using a systems value-stream structure. This uses a proactive and problem-preempting approach to execute projects. This alternative has the following advantages:

- Problem diagnosis is easier
- Accountability is higher
- Operating waste is minimized
- Conflict resolution is faster
- Value points are traceable

SYSTEMS VALUE MODELING

A technique that can be used to assess overall value-added components of a process improvement program is the systems value model (SVM). The model provides an analytical decision aid for comparing process alternatives. Value is represented as a p-dimensional vector:

$$V = f\left(A_1, A_2, \ldots, A_p\right)$$

where $A = (A_1, \ldots, A_n)$ is a vector of quantitative measures of tangible and intangible attributes. Examples of process attributes are quality, throughput, capability, productivity, cost, and schedule. Attributes are considered to be a combined function of factors, x_1, expressed as:

$$A_k\left(x_1, x_2, \ldots, x_{m_k}\right) = \sum_{i=1}^{m_k} f_i\left(x_i\right)$$

where $\{x_i\}$ = set of m factors associated with attribute $A_k(k = 1, 2, \ldots, p)$ and f_i = contribution function of factor x_i to attribute A_k. Examples of factors include reliability, flexibility, user acceptance, capacity utilization, safety, and design functionality. Factors are themselves considered to be composed of indicators, v_i, expressed as

$$x_i\left(v_1, v_2, \ldots, v_n\right) = \sum_{j=1}^{n} z_i\left(v_i\right)$$

where $\{v_j\}$ = set of n indicators associated with factor $x_i(i = 1, 2, ..., m)$ and z_j = scaling function for each indicator variable v_j. Examples of indicators are project responsiveness, lead time, learning curve, and work rejects. By combining the above definitions, a composite measure of the value of a process can be modeled as:

$$V = f\left(A_1, A_2, ..., A_p\right)$$

$$= f\left\{\begin{bmatrix}\sum_{i=1}^{m_1}f_1\left(\sum_{j=1}^{n}z_j\left(v_j\right)\right)\end{bmatrix}_1, \begin{bmatrix}\sum_{i=1}^{m_2}f_2\left(\sum_{j=1}^{n}z_j\left(v_j\right)\right)\end{bmatrix}_2, ..., \begin{bmatrix}\sum_{i=1}^{m_k}f_p\left(\sum_{j=1}^{n}z_j\left(v_j\right)\right)\end{bmatrix}_p\right\}$$

where m and n may assume different values for each attribute. A subjective measure to indicate the utility of the decision-maker may be included in the model by using an attribute weighting factor, w_i, to obtain a weighted PV:

$$PV_w = f\left(w_1 A_1, w_2 A_2, ..., w_p A_p\right)$$

where

$$\sum_{k=1}^{p}w_k = 1, \qquad \left(0 \le w_k \le 1\right)$$

With this modeling approach, a set of process options can be compared on the basis of a set of attributes and factors.

To illustrate the model above, suppose three IT options are to be evaluated based on four attribute elements: *capability, suitability, performance,* and *productivity*. For this example, based on the equations, the value vector is defined as:

$$V = f\left(capability, \; suitability, \; performance, \; productivity\right)$$

Capability: The term "capability" refers to the ability of IT equipment to satisfy multiple requirements. For example, a certain piece of IT equipment may only provide computational service. A different piece of equipment may be capable of generating reports in addition to computational analysis, thus increasing the service variety that can be obtained. In the analysis, the levels of increase in service variety from the three competing equipment types are 38 percent, 40 percent, and 33 percent, respectively. *Suitability*: "Suitability" refers to the appropriateness of the IT equipment for current operations. For example, the respective percentages of operating scope for which the three options are suitable for are 12 percent, 30 percent, and 53 percent. *Performance*: "Performance," in this context, refers to the ability of the IT equipment to satisfy schedule and cost requirements. In the example, the three options can, respectively, satisfy requirements on 18 percent, 28 percent, and 52 percent of the typical set of jobs. *Productivity*: "Productivity" can be measured by an assessment of the performance of the proposed IT equipment to meet workload requirements in relation to the existing equipment. For the example, the three options, respectively, show normalized increases of 0.02, −1.0, and −1.1 on a uniform scale of productivity measurement. Option C is the best "value" alternative in terms of suitability and performance. Option B shows the best capability measure, but its productivity is too low to justify the needed investment. Option A offers the best productivity, but its suitability measure is low. The analytical process can incorporate a lower control limit into the quantitative assessment such that any option providing value below that point will not be acceptable. Similarly, a minimum value target can be incorporated into the graphical plot such that each option is expected to exceed the target point on the value scale.

The relative weights used in many justification methodologies are based on subjective propositions of decision-makers. Some of those subjective weights can be enhanced by the incorporation of utility models. For example, the weights could be obtained from utility functions. There is a risk of spending too much time maximizing inputs at "point-of-sale" levels with little time defining and refining outputs at the "wholesale" systems level.

A systems view of operations is essential for every organization. Without a systems view, we cannot be sure we are pursuing the right outputs. A systems approach allows for a multi-dimensional analysis of any endeavor, considering many of the typical "ilities" of systems engineering as listed below:

- Affordability
- Practicality
- Desirability
- Configurability
- Modularity
- Reliability
- Desirability
- Maintainability
- Testability

- Transmittability
- Reachability
- Quality
- Agility

A systems engineering plan is essential for the following reasons:

1. Description of the system being developed
2. Description of team structure and responsibilities
3. Identification of all project stakeholders
4. Description of tailored technical activities in each phase
5. Documentation of decisions and technical implementation
6. Establishment of technical metrics and measurements (Who, What, When, Where, Which, How, Why)

Now that we have explained some of the characteristics of a system, we can move on to specific applications and other considerations in the chapters that follow.

MODEL-BASED SYSTEMS ENGINEERING

Industry 4.0 requires a robust and adaptive model of a system. A good model is a good guide for process execution. Model-Based Engineering (MBE) is a general approach to engineering that uses models as an integral part of the technical baseline that includes the requirements, analysis, design, implementation, and verification of a capability, system, and/or product throughout a systems life cycle. Model-Based Systems Engineering (MBSE) focuses MBE specifically on the approach of systems engineering. MBSE is the formalized application of modeling to support system requirements, design, analysis, verification and validation activities beginning in the conceptual design phase and continuing throughout development and later life cycle phases. MBSE provides significant opportunities for improved productivity, efficiency, effectiveness, and product quality. This is a relatively new approach to the application of systems engineering and has received growing attention and acceptance.

A model facilitates repeatability and consistency of activities in a systems environment. In general terms, a model is a simplified version of a concept, phenomenon, relationship, structure, organization, enterprise, or system. It can be a graphical, mathematical, flow diagram, or physical representation. A model is an abstraction of reality by focusing on the necessary components and eliminating or minimizing unnecessary components. The objectives of a representative model include the following:

- Facilitate understanding
- Aid in decision-making
- Examine 'what if' scenarios in a decision environment
- Explain the characteristics of a system
- Control events
- Predict events
- Describe systems profile
- Prescribe actions in the system
- Diagnose problems in a system

ELEMENTS OF MBSE

- Provides a repeatable template for systems actions
- Formalizes the practice of systems development through the use of models
- Broad scope with respect to multiple modeling domains across the lifecycle
- Guides both horizontal and vertical integration
- Facilitates improvements in quality and productivity
- Lowers the risk of operations
- Inherently embeds rigor and precision
- Enforces communication among internal and external teams and the customer
- Enhances the management of complexity

The replication of models horizontally and vertically provides a comprehensive view of the systems environment (see Badiru, 2019). In the relationship, the vertical flow goes from the component level to the system level and on to the operational level. The horizontal integration flows through the sequential stages. The MBSE interfaces and inner workings of the model are represented by the systems engineering elements. The overall framework essentially represents a meta-model structure, whereby each inner element is, itself, a representative model of a function or process. Model-based Systems Engineering provides a means for driving more systems engineering depth without increasing costs. Data-centric specifications within MBSE enable automation and optimization, and unity of focus on value-added tasks. It also ensures a balanced approach to functions as well as an increased level of systems understanding. Systems understanding can be achieved through integrated analytics that are tied to a model-centric technical baseline. MBSE also drives a consistent specification across the design spectrum. The key to a successful model-based approach to Industry 4.0 is scoping the problem. Pertinent questions include:

What is expected out of the model?
What level of fidelity is desired to be accomplished?
What are the success criteria for the application of the model?

Scoping and managing a modeling effort is both an art and a science. Since driving change in an organization takes time, interpersonal relationship, and commitment to continuous investment, the mix of art and science is very important. For a continuity of the model, the sustainment end point will loop back to the beginning for the purpose of identifying new needs for the next cycle. This is essential for Industry 4.0, where adaptability to changing technologies is critical. The characteristics of MBSE include:

- Set of interconnected models
- Models are an abstraction of reality
- Vetted structure
- Behavior and requirements
- Standard language
- Graphical notation
- Syntax
- Semantics
- Visual focus
- Static and dynamic components
- Shared system information base

In moving from document-centric approach to model-centric approach for Industry 4.0, the linkages are important. A good model will show a flow and interplay of various quantitative and qualitative aspects of operating the automobile system. Systems modeling helps accomplish the following:

- Improved system and software
  - Specification
  - Visualization

- Architecture
- Construction
- Simulation and Test
- Documentation
  - Validation and verification
- Improved communications
  - Enhanced knowledge capture and transfer
  - Training support
- Improved design quality
  - Decreased ambiguity
  - Increased precision
  - Supports evaluation of Consistency, Correctness, and Completeness
  - Supports evaluation of the trade space

Involvement of internal and external stakeholders is essential. The categories of stakeholders can be extensive, but basically include the following:

1. Customers
2. Project managers
3. Designers, developers, integrators
4. Vendors
5. Testers
6. Regulators

The impact of MBSE extend to several organizational aspects, including

- Systems architecture
- Cost
- Performance
- CAD (Computer Aided Design)
- Manufacturing
- Electronics
- Software
- Verification

With MBSE, decision-makers will have more information and options from which to draw conclusions. Integrated analytics models will both increase the amount of information available to decision makes and also help decision-makers make sense of the information. Tools to explore, visualize, and understand a complex trade space, rooted in MBSE, can provide early insight into the impact of decisions ranging from technical solutions to complex public policies.

The primary focus of most current industry efforts to move toward a Model-based Engineering approach focus on integrating data through models. By bringing together varied but related models into a data-rich, architecture-centric environment, new levels of systems understanding can be achieved. Model-based Systems Engineering forms a means to achieve integration.

SYSTEMS PROCESS IMPROVEMENT

Model-Based Systems call for the incorporation of models into the overall operational scheme of the organization. Over the past few decades, a process approach has come to dominate our view of how to conceptualize and organize work, from a systems perspective (Heminger, 2014). Current approaches to management, such as Business Process Reengineering (BPR), Lean, and Sixsigma are all based on this concept. Indeed, it seems almost axiomatic today to assume that this is the correct way to understand organizational work. Yet each of these approaches seems to say different things about processes. What do they have in common that supports using a process approach? And, what do their different approaches tell us about different types of problems with the management of organizational work? To answer these questions, it may help to take a historical look at how work has been done since before the industrial revolution up to today.

Prior to the industrial revolution, work was done largely by craftsmen, who underwent a process of becoming skilled in their trade of satisfying customers wants and needs. Typically, they started as apprentices, where they learned the rudiments of their craft from beginning to end, moved on to become journeymen, then craftsmen as they become more knowledgeable, finally, reaching the pinnacle of their craft as master craftsmen. They grew both in knowledge of their craft and in understanding what their customers wanted. In such an arrangement organizational complexity was low, with a few journeymen and apprentices working for a master craftsman. But, because work by craftsmen was slow and labor-intensive, only a few of the very wealthiest people could have their needs for goods met. Most people did not have access to the goods that the few at the top of the economic ladder were able to get. There was a long-standing and persistent unmet demand for more goods.

This unmet demand, coupled with a growing technological capability, provided the foundations for the industrial revolution. Manufacturers developed what Adam Smith (1776) called the "division of labor," in which complex tasks were broken down into simple tasks, automated where possible, and supervisors/managers were put in place to see that the pieces came together as a finished product. As we moved further into the industrial revolution, we continued to increase our productivity and the complexity of our factories. With the huge backlog of unmet demand, there was a willing customer for most of what was made. But, as we did this, an important change was taking place in how we made things. Instead of having a master craftsman in charge who knew both how to make goods as well as what the customers wanted and needed, we had factory supervisors, who learned how to make the various parts of the manufactured goods come together. Attention and focus began to turn inward from the customers to the process of monitoring and supervising complex factory work.

Over time, our factories became larger and ever more complex. More and more management attention needed to be focused inward on the issues of managing this complexity to turn out ever higher quantities of goods. In the early years of the twentieth century, Alfred Sloan, at General Motors, did for management what the industrial revolution had done for labor. He broke management down into small pieces, and assigned authority and responsibility tailored to those pieces. This allowed managers to focus on small segments of the larger organization, and to manage according to the authority and respon-sibility assigned. Through this method, General Motors was able to further advance productivity in the workplace. Drucker (1993) credits this internal focus on improved productivity for the creation of the middle class over the past one hundred years. Again, because of the long-standing unmet demand, the operative concept was that if you could make it, you could sell it. The ability to turn out huge quantities of goods culminated in the vast quantities of goods created in the United States during and immediately following World War II. This was added to by manufacturers in other countries, which came back on line after having their factories damaged or destroyed by the effects of the war. As they rebuilt and began producing again, they added to the total quantities of goods being produced.

Then, something happened that changed everything. Supply started to out-strip demand. It didn't happen everywhere evenly, either geographically, or by industry. But, in ever-increasing occurrences, factories found themselves sup-plying more than people were demanding. We had reached a tipping point. We went from a world where demand outpaced supply to a world where, increas-ingly, supply outpaced demand (Hammer and Champy, 1993). Not everything being made was going to sell; at least not for a profit. When supply outstrips demand, customers can choose. And, when customers can choose, they will choose. Suddenly, manufacturers were faced with what Hammer and Champy call the "3 Cs": customers, competition and change (Hammer and Champy, 1993). Customers were choosing among competing products, in a world of constant technological change. To remain in business, it was now necessary to produce those products that customers will choose. This required knowing what customers wanted. But management and the structure of organizations from the beginning of the industrial revolution had been largely focused inward, on raising productivity and making more goods for sale. Managerial structure, information flows, and decision points were largely designed to support the efficient manufacturing of more goods, not on tailoring productivity to the needs of choosy customers.

BUSINESS PROCESS REENGINEERING

A concept was needed that would help organizations focus on their customers and their customers' needs. The development of Business Process Reengineering (BPR) originated from that need. A process view of work provided a path for refocusing organizational efforts on meeting customer needs and expectations. On one level, a process is simply a series of steps, taken in some order, to achieve some result. Hammer and Champy (1993), however, provided an important distinction in their definition of a process. They defined it as "a collection of activities that takes one or more inputs and creates an output that is of value to the customer." By adding the customer to the definition, Hammer and Champy provided a focus back on the customer, where it had been prior to the industrial revolution. In their 1993 book, *Reengineering the Corporation: A Manifesto for Business Revolution*, Hammer and Champy advocated BPR, which they defined as "the fundamental rethinking and radical redesign of business processes to achieve dramatic improvements in critical, contemporary measures of performance…" In that definition, they identified four words that they believed were critical to their understanding of reengineering. Those four words were "fundamental," "radical," "dramatic," and "processes." In the following editions of their book, which came out in 2001 and 2003, they revisited this definition, and decided that the key word underlying all of their efforts was the word "process." And, with process defined as "taking inputs, and turning them into outputs of value to a customer," customers and what customers' value are the focus of their approach to reengineering.

Hammer and Champy viewed BPR as a means to rethink and redesign organizations to better satisfy their customers. BPR would entail challenging the assumption under which the organization had been operating, and redesigning around their core processes. They viewed the creative use of information technology as an enabler that would allow them to provide the information capabilities necessary to support their processes while minimizing their functional organizational structure.

LEAN PRINCIPLES FOR SYSTEMS 4.0

At roughly the same time that this was being written by Hammer and Champy, Toyota was experiencing increasing success and buyer satisfaction through its use of Lean, which is a process view of work focused on removing waste from the value stream. Womack and Jones (2003) identified the first of the Lean principles as value. And, they state, "Value can only be defined by the ultimate customer." So, once again, we see a management concept that leads organizations back to focus on their customers. Lean is all about identifying waste in a value stream (similar to Hammer and Champy's process) and removing that waste wherever possible. But the identification of what is waste can only be determined by what does or does not contribute to value, and value can only be determined by the ultimate customer. So, once again, we have a management approach that refocuses organizational work on the customers and their values.

Lean focuses on five basic concepts: value, the value stream, flow, pull, and perfection. "Value," which is determined by the ultimate customer, and the "value stream" can be seen as similar to Hammer and Champy's "process," which focuses on adding value to its customers. "Flow" addresses the passage of items through the value stream, and it strives to maximize the flow of quality production. "Pull" is unique to Lean and is related to the "just-in-time" nature of current manufacturing. It strives to reduce in-process inventory that is often found in large manufacturing operations. "Perfection" is the goal that drives Lean. It is something to be sought after, but never to be achieved. Thus, perfection provides the impetus for constant process improvement.

SIX SIGMA METHODOLOGY FOR SYSTEMS 4.0

In statistical modeling of manufacturing processes sigma refers to the number of defects per given number of items created. Six Sigma refers to a statistical expectation of 3.4 defects per million items. General Electric adopted this concept in the development of the Six Sigma management strategy in 1986. While statistical process control can be at the heart of a Six Sigma program, General Electric and others have broadened its use to include other types of error reduction as well. In essence, Six Sigma is a program focused on reducing errors and defects in an organization. While Six Sigma does not explicitly refer back to the customer for its source of creating quality, it does address the concept of reducing errors and variations in specifications. Specifications can be seen as coming from customer requirements; so again, the customer becomes key to success in a Six Sigma environment.

Six Sigma makes the assertion that quality is achieved through continuous efforts to reduce variation in process outputs. It is based on collecting and analyzing data, rather than depending on hunches or guesses as a basis for making decisions. It uses the steps define, measure, analyze, improve, and control (DMAIC) to improve existing processes. To create new processes, it uses the steps define, measure, analyze, design, and verify (DMADV). Unique to this process improvement methodology, Six Sigma uses a series of karate-like levels (yellow belts, green belts, black belts, and master black belts) to rate practitioners of the concepts in organizations. Many companies who use Six Sigma have been satisfied by the improvements that they have achieved. To the extent that output variability is an issue for quality, it appears that Six Sigma can be a useful path for improving quality.

From the above descriptions, it is clear that while each of these approaches uses a process perspective, they address different problem sets, and they suggest different remedies. BPR addresses the problem of getting a good process for the task at hand. It recognizes that many business processes over the years have been designed with an internal focus, and it uses a focus on the customer as a basis for redesigning processes that explicitly address what customers need and care about. This approach would make sense where organizational processes have become focused on internal management needs, or some other issues, rather than on the needs of the customer.

The Lean methodology came out of the automotive world, and is focused on gaining efficiencies in manufacturing. Although it allows for redesigning brand new processes, its focus appears to be most focused on working with an existing assembly line and finding ways to reduce its inefficiencies. This approach would make sense for organizations which have established processes/value streams where there is a goal to make those processes/value streams more efficient.

Six Sigma was developed from a perspective of statistical control of industrial processes. At its heart, it focuses on variability in processes and error rates in production and seeks to control and limit variability and errors where possible. It asserts that variability and errors cost a company money, and that learning to reduce these will increase profits. Similar to both BPR and Lean, it is dependent on top level support to make the changes that will provide its benefits.

Whichever of these methods is selected to provide a more effective and efficient approach to doing business, it may be important to remember the lessons of the history of work since the beginning of the industrial revolution. We started with craftsmen satisfying the needs of a small base of customers. We then learned to increase productivity to satisfy the unmet demand of a much larger customer base, but in organizations that were focused inward on issues of productivity, not outward toward the customers. Now that we have reached a tipping point where supply can overtake demand, we need to again pay attention to customer needs for our organizations to survive and prosper. One of the process views of work may provide the means to do that.

CONCLUSIONS

Chapter 1 covers the fundamentals of Industry 4.0. Chapter 2 covers the basic principles of systems engineering. Enmeshing both methodologies leads to the composite notion of Systems 4.0 for Industry 4.0. Systems 4.0 is the contemporary alignment of systems framework to the stages of the industrial revolution. As the industrial revolution has changed and advanced over the generations, so have the tenets of systems thinking. In this context, Systems 4.0 leverages the emergence of new technological assets, particularly intelligent software tools and smart hardware assets.

REFERENCES

ANSI/EIA-632-1988 (1999), *Processes for Engineering a System*, New York, NY: American National Standards Institute.

ANSI/EIA-731.1 (2002), *Systems Engineering Capability Model*, New York, NY: American National Standards Institute.

Badiru, Adedeji B. (2019), *Systems Engineering Models: Theory, Methods, and Applications*, Taylor & Francis/CRC Press, Boca Raton, FL.

Boehm, B.W. (1981), *Software Engineering Economics*, Prentice-Hall, Upper Saddle River.

Boehm, B.W. (1994), "Integrating software engineering and systems engineering," *The Journal of NCOSE 1*(1): 147–151.

Boehm, B.W., Abts, C., Brown, A.W., Chulani, S., Clark, B., Horowitz, E., Madachy, R., Reifer, D.J., and Steece, B. (2000), *Software cost estimation with COCOMO II*, Prentice-Hall, Upper Saddle River.

Boehm, B.W., Valerdi, R., and Honour, E. (2008), "The ROI of systems engineering: some quantitative results for software-intensive systems," *Systems Engineering 11*(3): 221–234.

Boehm, B.W., Valerdi, R., Lane, J., and Brown, A. W. (2005), "COCOMO suite methodology and evolution," *CrossTalk – The Journal of Defense Software Engineering 18*(4): 20–25.

CMMI, (2002), *Capability Maturity Model Integration - CMMI-SE/SW/IPPD/SS, V1.1*, Pittsburgh, PA, Carnegie Mellon – Software Engineering Institute.

Colombi, John, and Richard Cobb (2009), "Defense Acquisition Research Journal." *Application of Systems Engineering to Rapid Prototyping for Close Air Support*, 284–303.

DAU (2017), "Systems Engineering Process." Defense Acquisition University, Defense Acquisition University, September 29, 2017, www.dau.mil/acquipedia/Pages/ArticleDetails.aspx?aid=9c591ad6-8f69-49dd-a61d-4096e7b3086c.

Drucker, Peter F. (1993), *The Post Capitalist Society*, Harper Collins, New York, NY.

Furness, Dyllan (2018), "From BigDog to SpotMini: Tracing the Evolution of Boston Dynamics Robo-Dogs." *Digital Trends*, June 12. www.digitaltrends.com/cool-tech/evolution-boston-dynamics-spot-mini/.

Gerstner, Louis (2002), *Who Says Elephants Can't Dance?* Harper Collins, New York, NY.

Haas, R, et al. (2009), "A Case Study on Applying the Systems Engineering Approach: Best Practices and Lessons Learned from the Chattanooga SmartBus Project," *Home – Transport Research International Documentation - TRID*, October, 31, 2009, trid.trb.org/view.aspx?id=920390.

Hammer, M. and Champy, J. (1993, 2001, 2003), *Reengineering the Corporation: A Manifesto for Business Revolution*, Harper Business, New York.

Heminger, Alan (2014), "Industrial Revolution, Customers, and Process Improvement," in *Handbook of Industrial and Systems Engineering*, A. B. Badiru, editor, CRC Press/Taylor & Francis, Boca Raton, FL.

Henning, Wade A., and Walter, Daniel T. (2012), "Spiral Development in Action: A Case Study of Spiral Development in the Global Hawk Unmanned Aerial Vehicle Program", *Naval Postgraduate School*, December 2012. calhoun.nps.edu/bitstream/handle/10945/9980/05Dec_Henning_MBA.pdf?sequence=1.

ISO/IEC 15288:2002(E) (2002), *Systems Engineering – System Life Cycle Processes*, 1st Edition, International Organization for Standardization (ISO), Geneva, Switzerland, https://en.wikipedia.org/wiki/ISO/IEC_15288

Jia-Ching, Lin (2011), "Various Approaches for Systems Analysis and Design." *Populations and Sampling*, August, 11, 2011, www.umsl.edu/~sauterv/analysis/termpapers/f11/jia.html.

London, Brian (2012), *A Model-Based Systems Engineering Framework for Concept Development*. PDF. MIT, Cambridge, March 8, 2012.

McMurtry, Mark (2013), "A Case Study of the Application of the Systems Development Life Cycle (SDLC) in 21st Century Health Care: Something Old, Something New?" *Papers of the Abraham Lincoln Association, Michigan Publishing*, University of Michigan Library, January, 1, 2013, https://www.researchgate.net/publication/269662156_A_Case_Study_of_the_Application_of_the_Systems_Development_Life_Cycle_SDLC_in_21st_Century_Health_Care_Something_Old_Something_New

MIL-STD 490-A (1985), *Specification Practices*, U.S. Department of Defense, Washington, DC.

MIL-STD-499A (1969), *Engineering Management* U.S. Department of Defense, Washington, DC.

Narasimhan, K. (2002), "The Six Sigma Way: How GE, Motorola, and Other Top Companies are Honing Their Performance," https://www.emerald.com/insight/content/doi/10.1108/tqmm.2002.14.4.263.1/full/html

Pande, P. S., R. P. Neuman, and Cavanaugh, R.R. (2000), *The Six Sigma Way: How to Maximize the Impact of Your Change and Improvement Efforts*, McGraw-Hill, New York, NY.

Pandikow, A. and Törne, A., "Integrating Modern Software Engineering and Systems Engineering Specification Techniques," *14th International Conference on Software & Systems Engineering and their Applications*, Vol. 2, 2001.

Richtin, Eberhardt (1999), *Systems Architecting of Organizations: Why Eagles Can't Swim*, CRC Press, Boca Raton, FL.

Satzinger, John W. et al. (2016), "Approaches to Systems Development," in *Systems Analysis and Design in a Changing World*, 6th ed., Cengage Learning, pp. 81–89.

Smith, A., *The Wealth of Nations*, (orig. 1776, 2012) Simon & Brown Publishers, New York, NY.

Tutorials Point (2018), "V-Model." www.tutorialspoint.com. July 21, 2018. Accessed August 8, 2018. https://www.tutorialspoint.com/sdlc/sdlc_v_model.htm

Valerdi, R. (2014), "Systems engineering cost estimation with a parametric model," in *Handbook of Industrial and Systems Engineering*, A. B. Badiru, Editor, CRC Press, Boca Raton, FL.

Valerdi, R., Wheaton, M. J., and Fortune, J. (2007), "Systems Engineering Cost Estimation for Space Systems," in *AIAA Space*, AIAA 2007-6001, Los Angeles, CA.

Valerdi, R. (2005), The Constructive Systems Engineering Cost Estimation Model (COSYSMO), PhD Dissertation, University of Southern California, 2005.

Womack, J. P. and Jones, D. T., (2003), *Lean Thinking: Banish Waste and Create Wealth in Your Corporation*, New York: Free Press.

Digital Manufacturing in Industry 4.0

3

ELEMENTS OF DIGITAL MANUFACTURING

Digital manufacturing (aka 3D Printing, Additive Manufacturing) is one of the driving technologies for Industry 4.0. This is covered in this chapter, along with some traditional manufacturing processes, to show the attendant transformation over the years. Industry 4.0 is revolutionizing the way companies manufacture, improve and distribute their products. Manufacturers are integrating new technologies, including 3D Printing (additive manufacturing), Internet of Things (IoT), cloud computing and analytics, and AI and machine learning into their production facilities and throughout their operations. These smart factories are equipped with advanced sensors, embedded software and robotics that collect and analyze data and allow for better decision-making. Even higher value is created when data from production operations is combined with operational data from ERP, supply chain, customer service and other enterprise systems to create whole new levels of visibility and insight from previously siloed information.

Digital technologies lead to increased automation, predictive maintenance, the self-optimization of process improvements and, above all, a new level of efficiencies and responsiveness to customers not previously possible. Developing smart factories provides an incredible opportunity for the manufacturing industry to enter the fourth industrial revolution. Analyzing the

DOI: 10.1201/9781003312277-3

large amounts of big data collected from sensors on the factory floor ensures real-time visibility of manufacturing assets and can provide tools for performing predictive maintenance in order to minimize equipment downtime. Using high-tech IoT devices in smart factories leads to higher productivity and improved quality. Replacing manual inspection business models with AI-powered visual insights reduces manufacturing errors and saves money and time. With minimal investment, quality control personnel can set up a smartphone connected to the cloud to monitor manufacturing processes from virtually anywhere. By applying machine learning algorithms, manufacturers can detect errors immediately, rather than at later stages when repair work is more expensive.

Industry 4.0 concepts and technologies can be applied diverse manufacturing systems, including discrete and process manufacturing, as well as the oil and gas industry, mining, and other industrial enterprises.

From a technological standpoint (Raman and Wadke, 2014), manufacturing involves the making of products from raw materials through the use of human labor and resources which include machines, tools and facilities. It could be more generally regarded as the conversion of an unusable state into a usable state by adding value along the way. For instance, a log of wood serves as the raw material for making lumber, which, in turn, is the raw material to produce chairs. The value added is usually represented in terms of cost and/or time. The term "manufacturing" originates from the Latin word *manufactus*, which means "made by hand." Manufacturing has seen several advances during the last three centuries: mechanization, automation, and, most recently, computerization, leading to the emergence of direct digital manufacturing, which is popularly known as 3D printing.

Processes that used to be predominantly done by hand and hand tools have evolved into sophisticated processes making use of cutting-edge technology and machinery. A steady improvement in quality has resulted, with today's specifications even on simple toys exceeding those achievable just a few years ago. Mass production, a concept developed by Henry Ford, has advanced so much that it is now a complex, highly agile, and highly automated manufacturing enterprise. There are several managerial and technical aspects of embracing new technologies to advance manufacturing. Sieger and Badiru (1993), Badiru (1989, 1990, 2005), and others address the emergence and leveraging of past manufacturing-centric techniques of expert systems, flexible manufacturing systems (FMS), nano-manufacturing, and artificial neural networks. In each case, strategic implementation, beyond hype and fad, is essential for securing the much-touted long-term benefits. It is envisioned that the contents of this handbook will expand the knowledge of readers to facilitate the strategic embrace of 3D printing.

LEVERAGING 3D PRINTING

3D printing, also known as Additive Manufacturing or Direct Digital Manufacturing, is a process for making a physical object from a three-dimensional digital model, typically by laying down many successive thin layers of a material. The successive layering of materials constitutes the technique of additive manufacturing. Thus, the term "direct digital manufacturing" stems from the process of going from a digital blueprint of a product to a finished physical product. Manufacturers can use 3D printing to make prototypes of products before going for full production. In educational settings, faculty and students use this process to make project-related prototypes. Open-source and consumer-level 3D printers allow for creating products at home, thus advancing the concept of distributed additive manufacturing. Defense-oriented and aerospace products are particularly feasible for the application of 3D printing. The military often operate in remote regions of the world, where quick replacement of parts may be difficult to accomplish. With 3D printing, rapid production of routine replacement parts can be achieved at low cost onsite to meet urgent needs. The military civil engineering community is particularly fertile for the application of 3D printing for military asset management purposes.

By all accounts, 3D printing is now energizing the world of manufacturing. The concept of 3D printing was initially developed by Charles W. Hull in the 1980s (US Patent, 1986) as a stereolithography tool for making basic polymer objects. Today, the process is used to make intricate aircraft and automobile components. We are now seeing more and more applications in making prostheses. The first commercial 3D printing product came out in 1988 and proved a hit among auto manufacturers and aerospace companies. The design of medical equipment has also enjoyed a boost due to 3D printing capabilities. The possibilities appear endless from home-printed implements to the printing of complex parts in outer space. The patent abstract (US Patent, 1986) for 3D printing states the following:

A system for generating three-dimensional objects by creating a cross-sectional pattern of the object to be formed at a selected surface of a fluid medium capable of altering its physical state in response to appropriate synergistic stimulation by impinging radiation, particle bombardment or chemical reaction, successive adjacent laminae, representing corresponding successive adjacent cross-sections of the object, being automatically formed and integrated together to provide a step-wise laminar buildup of the desired object, whereby a three-dimensional object is formed and drawn from a substantially planar surface of the fluid medium during the forming process.

In order to understand and appreciate the full impact and implications of direct digital manufacturing (3D printing) we need to understand the traditional manufacturing processes that 3D printing is rapidly replacing. The sections that follow are reprinted (with permission) from Sirinterlikci (2014).

NATIONAL IMPACTS OF MANUFACTURING

Hitomi (1996) differentiated between the terms "production" and "manufacturing". According to him, production encompasses both making tangible products and providing intangible services while manufacturing is the transformation of raw materials into tangible products. Manufacturing is driven by a series of energy applications, each of which causes well-defined changes in the physical and chemical characteristics of the materials (Dano, 1966).

Manufacturing has a history of several thousand years and may impact humans and their nations in the following ways (Hitomi, 1994):

- *Providing basic means for human existence*: Without manufacture of products and goods humans are unable to live, and this is becoming more and more critical in our modern society.
- *Creating wealth of nations*: The wealth of a nation is impacted greatly by manufacturing. A country with a diminished manufacturing sector becomes poor and cannot provide a desired high standard of living to its people.
- *Moving toward human happiness and stronger world's peace*: Prosperous countries can provide better welfare and happiness to their people in addition to stronger security while posing less of a threat to their neighbors and each other.

In 1991, the National Academy of Engineering/Sciences in Washington, D.C. rated manufacturing as one of the three critical areas necessary for America's economic growth and national security, the others being science and technology (Hitomi, 1996). In recent history, nations which became active in lower-level manufacturing activities have grown into higher-level advanced manufacturing and a stronger research standing in the world (Gallagher, 2012).

As the raw materials are converted into tangible products by manufacturing activities, the original value (monetary worth) of the raw materials is increased (Kalpakjian and Schmid, 2006). Thus, a wire coat hanger has a greater value than its raw material, the wire. Manufacturing activities may produce *discrete products* such as engine components, fasteners, gears, or *continuous products* like sheet metal, plastic tubing, conductors which are later used in making of discrete products. Manufacturing occurs in a complex environment that connects multiple other activities: product design, process planning and tool engineering, materials engineering, purchasing and receiving, production control, marketing and sales, shipping, customer and support services (Kalpakjian and Schmid, 2006).

DIVERSITY OF MANUFACTURING PROCESSES

Today's manufacturing processes are extensive and continuously expanding while presenting multiple choices for manufacturing a single part of a given material (Kalpakjian and Schmid, 2006). The processes can be classified as traditional and non-traditional before they can be divided into their mostly physics-based categories. While much of the traditional processes have been around for a long time, some of the non-traditional processes may have been in existence for some time as well, such as in the case of electro-discharge machining, but not utilized as a controlled manufacturing method until a few decades ago.

Traditional manufacturing processes can be categorized as:

1. Casting and molding processes
2. Bulk and sheet forming processes
3. Polymer processing
4. Machining processes
5. Joining processes
6. Finishing processes

Non-traditional processes include:

1. Electrically-based machining
2. Laser machining
3. Ultrasonic welding
4. Water-jet cutting
5. Powder metallurgy
6. Small-scale manufacturing
7. Additive manufacturing
8. Bio-manufacturing

PROCESS PLANNING AND DESIGN

Selection of a manufacturing process or a sequence of processes depends on a variety of factors, including the desired shape of a part and its material properties for performance expectations (Kalpakjian and Schmid, 2006). Mechanical properties such as strength, toughness, ductility, hardness, elasticity, fatigue, and creep; physical properties such as density, specific heat, thermal expansion and conductivity, melting point, magnetic and electrical properties as well as chemical properties such as oxidation, corrosion, general degradation, toxicity, and flammability may play a major role in the duration of the service life of a part and recyclability. The manufacturing properties of materials are also critical since they determine whether the material can be cast, deformed, machined, or heat treated into the desired shape. For example, brittle and hard materials cannot be deformed without failure or high energy requirements; whereas they cannot be machined unless a non-traditional method such as electro-discharge machining is employed. Each manufacturing process has its characteristics, advantages and constraints including production rates and costs. For example, the conventional blanking and piercing process used in making sheet metal parts can be replaced by its laser-based counterparts if the production rates and costs can justify such a switch. Eliminating the need for tooling will also be a plus as long as the surfaces delivered by the laser-cutting process is comparable or better than that of the conventional method (Kalpakjian and Schmid, 2006). Quality is a subjective metric in general. However, in manufacturing, it often implies *surface finish and tolerances, both dimensional and geometric.* The economics of any process is again very important and can be conveniently decomposed with the analysis of manufacturing operations and their tasks. A manufactured part can be broken into its features, the features can be meshed with certain operations, and operations can be separated into their tasks. Since several possible operations may be available and multiple sequences of operations co-exist, several viable process plans can be made (Raman and Wadke, 2006).

Process routes are a sequence of operations through which raw materials are converted into parts and products. They must be determined after completion of production planning and product design according to the conventional wisdom (Hitomi, 1996). However, newer concepts like Concurrent or Simultaneous Engineering or Design for Manufacture and Assembly (DFMA) are encouraging simultaneous execution of part and process design and planning processes and additional manufacturing related activities. Process planning includes the two basic steps (Timms and Pohlen, 1970):

1. *Process Design* is a macroscopic decision-making for an overall process route for the manufacturing activity.
2. *Operation/Task Design* is a microscopic decision-making for individual operations and their detail tasks within the process route.

The main problems in process and operation design are: analysis of the workflow (flow-line analysis) for the manufacturing activity, and selecting the workstations for each operation within the workflow (Hitomi, 1996). These two problems are interrelated and must be resolved at the same time. If the problem to be solved is for an existing plant, the decision is made within the capabilities of that plant. On the contrary, an optimum workflow is determined, and then the individual workstations are developed for a new plant within the financial and physical constraints of the manufacturing enterprise (Hitomi, 1996).

Workflow is a sequence of operations for manufacturing activity. It is determined by manufacturing technologies, and forms the basis for operation design and layout planning. Before an analysis of workflow is completed, certain factors have to be defined, including precedence relationships and workflow patterns. There are two possible relationships between any two operations of the workflow (Hitomi, 1996):

1. A partial order, *precedence*, exists between two operations such as in the case of counterboring. Counterboring must be conducted after drilling.
2. No precedence exists between two operations if they can be performed in parallel or concurrently. Two sets of holes with different sizes in a part can be made in any sequence or concurrently.

Harrington (1973) identifies three different workflow patterns: *sequential* (tandem) process pattern of gear manufacturing; *disjunctive* (decomposing) pattern of coal or oil refinery processes; and *combinative* (synthesizing) process pattern in assembly processes.

According to Hitomi (1996), there are several alternatives for workflow analysis depending on production quantity (demand volume, economic lot size), existing production capacity (available technologies, degree of automation), product quality (surface finish, dimensional accuracy and tolerances), raw materials (material properties, manufacturability). The best workflow is selected by evaluating each alternative based on a criterion that minimizes *the total production (throughput) time*, or *total production cost*, *Operation process or flow process charts* can be used to define and present information for the workflow of the manufacturing activity. Once an optimum workflow is determined, the detail design process of each operation and its tasks are conducted. A *break-even analysis* may be needed to select the right equipment for the workstation. Additional tools, such as *man-machine analysis* as well as *human factors analysis*, are also used to define the details of each operation. *Operation sheets* are another type of tool used to communicate about the requirements of each task making up individual operations.

$$\text{total production time} = \Sigma \left[\text{transfer time between stages} + \text{waiting time} \right.$$
$$\left. + \text{set-up time} + \text{operation time} + \text{inspection time} \right]$$

$$\text{total production cost} = \text{material cost} + \Sigma \left[\text{cost of transfer between stages} \right.$$
$$+ \text{set-up cost} + \text{operation cost} + \text{tooling cost}$$
$$\left. + \text{inspection cost} + \text{work-in-process inventory cost} \right]$$

where Σ represent all stages of the manufacturing activity

Industrial engineering and operations management tools have been used to determine optimum paths for the workflow in manufacturing (Baudin and Netland, 2023). Considering the amount of effort involved for the complex structure of today's manufacturing activities, Computer aided process planning (CAPP) systems have become very attractive in order to generate feasible sequences and to minimize the lead time and non-value-added costs (Raman and Wadke, 2006).

DIGITIZATION PROCESSES

In order to digitize manufacturing, we must understand the traditional or conventional manufacturing processes to be replaced by 3D printing. For this reason, selected manufacturing processes are presented. Casting and molding processes can be classified into four categories (Kalpakjian and Schmid, 2006):

1. Permanent mold-based: Permanent molding, (high and low pressure) die-casting, centrifugal casting, and squeeze casting.

2. Expandable mold and permanent pattern-based: Sand casting, shell mold casting, and ceramic mold casting.
3. Expandable mold and expandable pattern: Investment casting, lost foam casting, and single-crystal investment casting.
4. Other processes: Melt-spinning.

These processes can be further classified based on their molds: permanent and expandable mold-type processes (Raman and Wadke, 2006). The basic concept behind these processes is to superheat a metal or metal alloy beyond its melting point or range, then pour or inject it into a die or mold, and allow it to solidify and cool within the tooling. Upon solidification and subsequent cooling, the part is removed from the tooling and finished accordingly. The expandable mold processes destroy the mold during the removal of the part or parts such as in sand casting and investment casting. Investment casting results in better surface finishes and tighter tolerances than sand casting. Die casting, and centrifugal casting processes also result in good finishes, but are permanent mold processes. In these processes, the preservation of tooling is a major concern since they are reused over and over, sometimes for hundreds of thousands of parts (Raman and Wadke, 2006). Thermal management of tooling through spraying and cooling channels is also imperative since thermal fatigue is a major failure mode for this type of tooling (Sirinterlikci, 2000).

Common materials that are cast include metals such as aluminum, magnesium, copper, and their low-melting-point alloys, including zinc alloys, cast iron, and steel (Raman and Wadke, 2006). The tooling used have simple ways of introducing liquid metal, feeding it into the cavity, and have mechanisms to exhaust air or gas entrapped within as well as to prevent defects such as shrinkage porosity and promote solid castings and easy removal of the part or parts. Cores and cooling channels are also included for making voids in the parts and controlled cooling of them to reduce cycle times for the process respectively. The die or mold design, metal fluidity, and solidification patterns are all critical to obtain high-quality castings. Suitable provisions are made through allowances to compensate for shrinkage and finishing (Raman and Wadke, 2006).

BULK FORMING PROCESSES

Forming processes include bulk-metal forming as well as sheet-metal operations. No matter what the type or nature of process is forming is mainly applicable to metals that are workable by plastic deformation. This constraint makes brittle materials not eligible for forming. Bulk forming is the combined application of temperature and pressure to modify shape of a solid object (Raman and Wadke, 2006). While cold-forming processes conducted near room temperature require higher pressures, hot working processes take advantage of the decrease in material strength. Consequent pressure and energy requirements are also much lower in hot working, especially when the material is heated above its recrystallization temperature, 60 percent of the melting point (Raman and Wadke, 2006). Net shape and near net shape processes accomplish part dimensions that are exact or close to specification requiring little or no secondary finishing operations. A group of operations can be included in the classification of bulk forming processes (Kalpakjian and Schmid, 2006):

1. Rolling processes: Flat rolling, shape rolling, ring rolling, and roll forging.
2. Forging processes: Open-die forging, closed-die forging, heading, and piercing.
3. Extrusion and drawing processes: Direct extrusion, cold extrusion, drawing, and tube drawing.

In the flat rolling process, two rolls rotating in opposite directions are utilized in reducing the thickness of a plate or sheet metal. This thickness reduction is compensated for by an increase in the length; when the thickness and width are similar, an increase in both width and length occurs based on the preservation of the volume of the parts (Raman and Wadke, 2006). A similar process, shape rolling is used for obtaining different shapes or cross-sections. Forging is used for shaping objects in a press and additional tooling, and may involve more than one pre-forming operation, including blocking, edging, and fullering (Raman and Wadke). Open-die forging is done on a flat anvil and closed-die forging process uses a die with a distinct cavity for shaping. Open-die forging is less accurate but can be used in making extremely large parts due to its ease on pressure and consequent power requirements. While mechanical hammers deliver sudden loads whereas hydraulic presses apply gradually increasing loads. Swaging is a rotary variation of the forging process, where a diameter of a wire is reduced by reciprocating movement of one or two opposing dies. Extrusion is the forcing of a billet out of a die opening similar to squeezing toothpaste out of its container, either directly or indirectly. This process enables fabrication of different cross-sections on long pieces (Raman and Wadke, 2006). In co-extrusion, two different materials are extruded simultaneously and bond with each other. On the contrary, drawing process is based on pulling of a material through an orifice to reduce the diameter of the material.

SHEET-FORMING PROCESSES

Stamping is a generic term used for sheet metal processes. They include processes, such as blanking, punching (piercing), bending, stretching, deep drawing, bending, and coining. Processes are executed singularly or consecutively to obtain complex sheet metal parts with a uniform sheet metal thickness. Progressive dies allow multiple operations to be performed at the same station. Since these tooling elements are dedicated, their costs are high and expected to perform without failure for the span of the production.

Sheet metal pieces are profiled by a number of processes based on a shear fracture of the sheet metal. In punching, a circular or a shaped hole is obtained by pushing the hardened die (punch) through the sheet metal. In a similar process called perforating, a group of punches are employed in making a hole pattern. In the blanking process, the aim is to keep the part that is punched out by the punch, not the rest of the sheet metal with the punched hole in the punching process. In the nibbling process, a sheet supported by an anvil is cut to a shape by successive bites of a punch similar to the motion of the sewing machine head.

After the perforation of a sheet metal part, multiple different stamping operations may be applied to it. In simple bending, the punch bends the blank on a die. Stretching may be accomplished while strictly holding the sheet metal piece by the pressure pads and forming it in a die whereas the sheet metal piece is allowed to be deeply drawn into the die while being held by the pressure pads in deep drawing. More sophisticated geometries can be obtained by roll forming in a progressive setting.

POLYMER PROCESSES

Polymer extrusion is a process to make semi-finished polymer products such as rods, tubes, sheets, and film in mass quantities. Raw materials such as pellets or beads are fed into the barrel to be heated and extruded at temperatures as high as 370 °C (698 °F). The extrusion is then air- or water-cooled and may later be drawn into smaller cross-sections. Variations of this process are film blowing, extrusion blow molding, and filament forming. This extrusion process is used in the making of blended polymer pellets and becomes a post-process for other processes such as injection molding. It is also utilized in coating metal wires in high speeds. Injection molding is a process similar to polymer extrusion with one main difference, the extrusion being forced into a metal mold for solidification under pressure and cooling. Feeding system into the mold includes the sprue area, runners, and gate. Thermoplastics, thermosets, elastomers, and even metal materials are being injection molded. Co-injection process allows molding of parts with different material properties, including colors and features. While injection foam molding with inert gas or chemical blowing agents results in the making of large parts with solid skin and cellular internal structure, reaction injection molding (RIM) mixes low-viscosity chemicals under low pressures (0.30–0.70 MPa) (43.51–101.53 psi) to be polymerized via a chemical reaction inside a mold. The RIM process can produce complex geometries and works with thermosets such as polyurethane or other polymers such as nylons, and epoxy resins. The RIM process is also adapted to fabricate fiber-reinforced composites.

Adapted from glass-blowing technology, the blow-molding process utilizes hot air to push the polymer against the mold walls to be frozen. The process has multiple variations, including extrusion and stretch blow molding. The generic blow-molding process allows the inclusion of solid handles and has better control over the wall thickness compared with its extrusion variant.

Thermoforming processes are used in making large sheet-based moldings. Vacuum thermoforming applies a vacuum to draw the heated and softened sheet into the mold surface to form the part. Drape thermoforming take advantages of the natural sagging of the heated sheet in addition to the vacuum whereas the plug-assisted variant of thermoforming supplements the vacuum with a plug by pressing on the sheet. In addition, pressure thermoforming applies a few atmospheres (atm) to push the heated sheet into the mold. A range of molding materials are employed in the thermoforming processes, including wood, metal, and polymer foam. A wide variety of other polymer-processing methods are available, including but not limited to rotational molding, compression molding, resin transfer molding (RTM).

MACHINING PROCESSES

Machining processes use a cutting tool to remove material from the work piece in the form of chips (Raman and Wadke, 2006). The cutting process requires plastic deformation and consequent fracture of the work-piece material. The type of chip impacts both the removal of the material and the quality of surface generated. The size and the type of the chip are a dependent of the type of machining operation and cutting parameters. The chip types are continuous, discontinuous, continuous with a built-up edge, and serrated (Raman and Wadke, 2006). The critical cutting parameters include the cutting speed (in revolutions per minute – rpms, surface feet per minute (sfpm) or millimeters per minute (mm/min)), the feed rate (inches per minute (ipm) or millimeters per minute (mm/min), and the depth of cut (inches (in) or millimeters (mm)). These parameters affect the work-piece, the tool, and the process itself (Raman and Wadke, 2006). The conditions of the forces, stresses, and temperatures of the cutting tool are determined by these parameters. Typically, the work-piece or tool are rotated or translated such that there is relative motion between the two. A primary zone of deformation causes a shear of material separating a chip from the work-piece. A secondary zone is also developed based on the friction between the chip and cutting tool (Raman and Wadke, 2006). While rough machining is an initial process to obtain the desired geometry without accurate dimensions and surface finish, finish machining is a precision process capable of great dimensional accuracy and surface finish. Besides metals, stones, bricks, wood, and plastics can be machined.

There are usually three types of chip-removal operations: single-point, multipoint (fixed geometry), and multipoint (random geometry) (Raman and Wadke, 2006). Random geometry multipoint operation is also referred to as an abrasive machining process, and includes operations such as grinding, honing, and lapping. The cutting tool in a single-point operation resembles a wedge, with several angles and radii to aid cutting. The cutting-tool geometry is characterized by the rake angle, the lead or main cutting-edge angle, the nose radius, and the edge radius. Common single-point operations include turning, boring, and facing (Raman and Wadke, 2006). Turning is performed to make round parts, facing makes flat features, and boring fabricates non-standard diameters, and internal cylindrical surfaces. Multipoint (fixed geometry) operations include milling and drilling. Milling operations can be categorized into face milling, peripheral (slab) milling, and end milling. The face milling uses the face of the tool, while slab milling uses the periphery of the cutter to generate the cutting action. These are typically applied to make flat features at a rate of material removal significantly higher than single-point operations like shaping and planning (Raman and Wadke, 2006). End milling cuts along with both the face and periphery and is used for making slots and extensive contours. Drilling is used to make standard-sized holes with a cutter with multiple active cutting edges (flutes). Rotary end of the cutter is used in the material removal process. Drilling has been the fastest and most economical method of making holes into a solid object. A multitude of drilling and relevant operations are available including core drilling, step (peck) drilling, counterboring, and countersinking, as well as reaming and tapping.

ASSEMBLY AND JOINING PROCESSES

Joining processes are employed in the manufacture of multi-piece parts and assemblies (Raman and Wadke, 2006).These processes encompass mechanical fastening through removable bolting shown in the illustration and non-removal riveting shown in the illustration, adhesive bonding, and welding processes. Welding processes use different heat sources to cause localized melting of the metal parts to be joined or the melting of a filler to develop a joint between mainly two metals – also being heated. Welding of plastics has also been established. Cleaned surfaces are joined together through a butt weld or a lap weld, although other configurations are also feasible (Raman and Wadke, 2006). Two other joining processes are brazing and soldering, which differ from each other in terms of the process temperatures and are not as strong as welding.

Arc welding utilizes an electric arc between two electrodes to generate the required heat for the process. One electrode is the plate to be joined while the other electrode is the consumable one. Stick welding, which is also called shielded metal arc welding (SMAW), is the most common. Metal inert gas (MIG), or gas metal arc welding, uses a consumable electrode (Raman and Wadke, 2006) as well. The electrode provides the filler and the inert gas provides an atmosphere such that contamination of the weld pool is prevented and consequent weld quality is obtained. A steady flow of electrode is accomplished through automatically to maintain the arc gap, sequentially controlling the temperature of the arc (Raman and Wadke, 2006). Gas tungsten arc welding or tungsten inert gas (TIG) welding uses a non-consumable electrode, and filler is required for the welding. In resistance welding, the resistance is generated by the air gap between the surfaces to obtain and maintain the flow of electric current between two fixed electrodes. The electrical current is then used to generate the heat required for welding (Raman and Wadke, 2006). This process results in spot or seam welds. Gas welding typically employs acetylene (fuel) and oxygen (catalyzer) to develop different temperatures to heat work-pieces or fillers for welding, brazing, and soldering. If the acetylene is in excess, a reducing (carbonizing) flame is obtained. The reducing flame is used in hard-facing or backhand pipe-welding operations. On the contrary, there is an excess of oxygen, then an oxidizing flame is generated. The oxidizing flame is used in braze-welding and welding of brasses or bronzes. Finally, if equal proportions of the two are used, a neutral flame results. The neutral flame is used in welding or cutting. Other solid-state processes include thermit welding, ultrasonic welding, and friction welding.

FINISHING PROCESSES

Finishing is often needed to complete the specialized work of 3D printing. Finishing processes include surface treatment processes and material removal processes such as polishing, shot-peening, sand blasting, cladding and electroplating, and coating and painting (Raman and Wadke, 2006). Polishing involves very little material removal and is also classified under machining operations. Shot and sand blasting are used to improve surface properties and cleanliness of parts. Chemical Vapor Deposition (CVD) or Physical Vapor Deposition (PVD) methods are also applied to improve surface properties (Kalpakjian and Schmid, 2006). Hard coatings such as CVD or PVD are applied to softer substrates to improve wear resistance while retaining fracture resistance (Raman and Wadke, 2006), as in the case metal coated polymer injection molding inserts (Noorani, 2006). The coatings are less than 10 mm thick in many instances. On the other hand, cladding is done as in the case of aluminum cladding on stainless steel to improve its heat conductivity for thermally critical applications (Raman and Wadke, 2006).

NON-TRADITIONAL MANUFACTURING PROCESSES

There are many non-traditional manufacturing processes, including:

1. Electrically-based machining: These processes include electro-discharge machining (EDM) and electro-chemical machining (ECM). In the Plunge EDM process, the work-piece is held in a work-holder submerged in a dielectric fluid. Rapid electric pulses are discharged between the graphite electrode and the work-piece, causing plasma to erode the work-piece. The dielectric fluid then carries the debris. The Wire EDM uses mainly brass wire in place of the graphite electrode, but functions in a similar way. The ECM process (also called reverse electroplating) is similar to the EDM processes, but it does not cause any tool wear, nor can any sparks be seen. Both processes can be used in machining very hard materials that are electrically conductive.

2. Laser machining: Lasers are used in a variety of applications ranging from cutting of complex 3D contours (i.e. today's coronary
stents) to etching or engraving patterns on rolls for making texture
on rolled parts. Lasers are also effectively used in hole-making, precision micro-machining, removal of coating, and ablation. Laser
transformation and shock-hardening processes make work-piece
surfaces very hard while the laser surface melting process produces
refined and homogenized microstructures.

3. Ultrasonic welding: Ultrasonic welding process requires an ultrasonic generator, a converter, a booster, and a welding tool. The generator converts 50 Hz into 20 KHz. These higher-frequency signals
are then transformed into mechanical oscillations through a reverse
piezoelectric effect. The booster and the welding tool transmit these
oscillations into the welding area, causing vibrations of 10–30 μm
in amplitude. Meanwhile a static pressure of 2–15 MPa is applied to
the work-pieces as they slide, get heated, and bonded.

4. Water-jet cutting: Water-jet cutting is an abrasive machining process employing abrasive slurry in a jet of water to machine hard
to machine materials (Raman and Wadke, 2006). Water is pumped
at high pressures like 400 MPa, and can reach speeds of 850 m/s
(3,060 km/hr or 1901.4 mph). Abrasive slurries are not needed when
cutting softer materials.

5. Powder metallurgy: Powder-based fabrication methods are critical
in employing materials with higher melting points due to the diffi-
culties in casting them. Once compressed under pressures using different methods and temperatures, compacted (green) powder parts
are sintered (fused) usually at two-thirds of their melting points.
Common powder metallurgy materials are ceramics and refractory
metals, stainless steel, and aluminum.

6. Small-scale manufacturing: The last two decades have seen a mar-
riage of micro-scale electronics device and their manufacturing
with mechanical systems – leading to the design and manufacturing of micro-electromechanical devices (MEMS). Even newer cutting-edge technologies have emerged on a smaller scale, such as
nanotechnology or molecular manufacturing (Raman and Wadke,
2006). The ability to modify and construct products at a molecular
level makes nanotechnology very attractive and usable. Single-wall
nanotubes are one of the biggest innovations for building future
transistors and sensors. Variants of the nano area include nano-manufacturing, such as ultra-high precision machining or adding ionic
antimicrobials to biomedical devices (Raman and Wadke, 2006)
(Sirinterlikci et al., 2012).

7. Additive manufacturing: Additive manufacturing has been driven from additive rapid prototyping technology. Since late 1980s rapid prototyping technologies have intrigued scientists and engineers. In the early years, there were attempts of obtaining 3D geometries using various layered approaches, as well as direct 3D geometry generation by robotic plasma spraying. Today, a few processes, such as Fused Deposition Modeling (FDM), Stereolithography (SLA), Laser Sintering Processes (Selective or Direct Metal), and 3D printing, have become familiar names. There are also other very promising processes, such as Objet's Polyjet (Inkjet) 3D printing technology. In the past two decades, the rapid prototyping technology has seen an increase in number of materials available for processing, and layer thicknesses has become less while control systems have improved to make the parts more accurate. The end result is shortened cycle times and better-quality functional parts. In addition, there have been many successful applications of rapid tooling and manufacturing.

8. Biomanufacturing: Biomanufacturing may encompass biological and biomedical applications. Thus, the manufacturing of human vaccinations may use biomass from plants while biofuels are extracted from corn or other crops. Hydraulic oils, printer ink technology, paints, cleaners and many other products are taking advantage of the developments in biomanufacturing (Sirinterlikci et al., 2010). On the other hand, biomanufacturing is working with nanotechnology, additive manufacturing, and other emerging technologies to improve the biomedical engineering field (Sirinterlikci et al., 2012).

EFFICACY OF MANUFACTURING SYSTEMS

The manufacturing systems are systems that are employed to manufacture products and the parts assembled into those parts (Groover, 2001). A manufacturing system is the collection of equipment, people, information, processes, and procedures to realize the manufacturing targets of an enterprise. Groover (2001) defines manufacturing systems in two parts:

1. Physical facilities include the factory, the equipment within the factory, and the way the equipment is arranged (the layout).
2. Manufacturing support systems is a set of procedures of the company to manage its manufacturing activities and to solve technical and logistics problems, including product design and some business functions.

Production quantity is a critical classifier of a manufacturing system while the way the system operates is another one, including *production to order* (i.e. Just in Time – JIT manufacturing) and *production for stock* (Hitomi, 1996). According to Groover (2001) and Hitomi 91996), a manufacturing system can be categorized into three groups based on its production volume:

1. Low production volume – Jobbing systems: In the range of 1–100 units per year, associated with a job shop, fixed position layout, or process layout.
2. Medium production volume – Intermittent or batch systems: With the production range of 100–10,000 units annually, associated with a process layout or cell.
3. High production volume – Mass manufacturing systems: 10,000 units to millions per year, associated with a process or product lay-out (flow line).

Each classification is associated inversely with a product variety level. Thus, when product variety is high, production quantity is low, and when product variety is low, production quantity is high (Groover, 2001). Hitomi (1996) states that only 15 percent of the manufacturing activities in the late 1990s were coming from mass manufacturing systems while small batch-multi product systems had a share more than 80 percent, perhaps due to diversification of human demands.

PART INTERCHANGEABILITY

While each manufacturing process is important for the fabrication of a single part, assembling these piece parts into subassemblies and assemblies also presents a major challenge. The concept that makes assembly a reality is interchangeability. Interchangeability relies on standardization of products and processes (Raman and Wadke, 2006). Besides facilitating easy assembly, interchangeability also enables easy and affordable replacement of parts within subassemblies and assemblies (Raman and Wadke, 2006). The key factors for interchangeability are "tolerances" and "allowances" (Raman and Wadke, 2006). Tolerance is the permissible variation of geometric features and part dimensions on manufactured parts with the understanding that perfect parts are hard to be made, especially repeatedly. Even if the parts could be manufactured perfectly, current measurement tools and systems may be unable to verify their dimensions and features accurately (Raman and Wadke, 2006). Allowances determine the degree of looseness or tightness of a fit in an assembly of two parts (i.e. a shaft and its bearing) (Raman and Wadke, 2006). Depending on the allowances, the fits are classified into "clearance," "interference," and "transition" fits. Since most commercial products and systems are based on assemblies, tolerances (dimensional or geometric) and allowances must be suitably specified to promote interchangeable manufacture (Raman and Wadke, 2006).

SYSTEMS METRICS

According to Hitomi (1996), efficient and economical execution of manufacturing activities can be achieved by completely integrating the material flow (manufacturing processes and assembly), information flow (manufacturing management system), and cost flow (manufacturing economics). A manufacturing enterprise need to serve for the welfare of the society by not harming the people and the environment (Green Manufacturing, Environmentally Conscious Manufacturing) along with targeting its profit objectives (Hitomi, 1996). A manufacturing enterprise needs to remain competitive and thus has to evaluate its products' values and/or the effectiveness of their manufacturing system by using the following four metrics (Hitomi, 1996):

1. Function and quality of products
2. Production costs and product prices
3. Production quantities (productivity) and on-time deliveries
4. Following any industry regulations

A variety of means are needed to support these metrics, including process planning and control, quality control, costing, safety, health and environment (SHE), production planning, scheduling, and control (including assurance of desired cycle time andtakt time and throughput times). Many other metrics are also used in the detail design and execution of systems, including machine utilization, floor space utilization, and inventory turnover rates.

AUTOMATION OF SYSTEMS

Some parts of a manufacturing system need to be automated while other parts remain under manual or clerical control (Groover, 2001). Either the actual physical manufacturing system or its support system can be automated as long as the cost of automation is justified. If both systems are automated, a high level of integration can also be reached as in the case of Computer-Integrated Manufacturing (CIM) Systems. Groover (2001) lists examples of automation in a manufacturing system:

- Automated machines
- Transfer lines that perform multiple operations
- Automated assembly operations
- Robotic manufacturing and assembly operations
- Automated material handling and storage systems
- Automated inspection stations

Automated manufacturing systems are classified into three groups (Groover, 2001):

1. Fixed Automation Systems: Used for high production rates or volumes and very little product variety, as in the case of a welding fixture used in making circular welds around pressure vessels.
2. Programmable Automation Systems: Used for batch production with small volumes and high variety, as in the case of a robotic welding cell.
3. Flexible Automation Systems: Medium production rates and varieties can be covered as in the case of flexible manufacturing cells with almost no lost time for changeover from one part family to another one.

Groover lists some of the reasons for automating a manufacturing entity:

1. To increase labor productivity and costs by substituting human workers with automated means
2. To mitigate skilled labor shortages as in welding and machining
3. To reduce routine and boring manual and clerical tasks
4. To improve worker safety by removing them from the point of operation in dangerous tasks such as nuclear, chemical, or high energy
5. To improve product quality and repeatability
6. To reduce manufacturing lead times
7. To accomplish processes that cannot be done manually
8. To avoid the high cost of not automating

CONCLUSIONS

Manufacturing is the livelihood and future of every nation and is mainly misunderstood. It also has a great role in driving engineering research and development; in addition, wealth generation. Utilization of efficient and effective methods is crucial for any manufacturing enterprise to remain competitive in this very global market, with intense international collaboration and rivalry. Especially the importance of industrial engineering tools in optimizing processes and integrating those processes into systems need to be grasped to better the manufacturing processes and their systems. Automation is still a valid medium for improving the manufacturing enterprise as long as the costs of doing it are justified. Tasks too difficult to automate, life cycles that are too short, products that are very customized, and cases where demands are very unpredictably varying, cannot justify the application of automation (Hitomi, 1996).

REFERENCES

Badiru, Adedeji B. (1989), "How to Plan and Manage Manufacturing Automation Projects," in *Proceedings of 1989 IIE Fall Conference*, Atlanta, GA, November 13–15, 1989, pp. 540–545.

Badiru, Adedeji B. (1990), "Analysis of Data Requirements for FMS Implementation is Crucial to Success," *Industrial Engineering 22*(10): 29–32.

Badiru, Adedeji B. (2005), *"Product Planning and Control for Nano-Manufacturing: Application of Project Management,"* Distinguished Lecture Series, University of Central Florida, Orlando, March 2005.

Baudin, Michel and Netland, Torbjorn (2023), *Introduction to Manufacturing: An Industrial Engineering and Management Perspective*, Routledge, New York, NY.

Dano, S. (1966), *Industrial Production Models: A Theoretical Study*, Springer, Vienna.

Gallagher, P. (2012), Presentation at the *National Network of Manufacturing Innovation Meeting II, Cuyahoga Community College*, Cleveland, OH, July 9, 2012.

Groover, M. (2001), *Automation, Production Systems, and Computer-Integrated Manufacturing*, Prentice-Hall, Upper Saddle River, NJ.

Harrington, J. Jr. (1973), *Computer Integrated Manufacturing*, Industrial Press, New York.

Hitomi, K. (1994), "Moving toward manufacturing excellence for future production perspectives," *Industrial Engineering 26*(6), p. 12

Hitomi, K. (1996), *Manufacturing Systems Engineering: A Unified Approach to Manufacturing Technology, Production Management, and Industrial Economics*, 2nd Edition, Taylor & Francis, Bristol, PA.

Kalpakjian, S., Schmid, S. R. (2006), *Manufacturing Engineering and Technology*, 5th Edition, Pearson Prentice-Hall, Upper Saddle River, NJ.

Noorani, R. (2006), *Rapid Prototyping and Applications*, John Wiley & Sons, Hoboken, NJ.

Raman, S. and Wadke, A. (2014), "Manufacturing Technology," in Badiru, A. B. (Ed), *Handbook of Industrial and Systems Engineering*, 2nd Edition, pp. 337–349, Taylor & Francis/CRC Press, Boca Raton, FL.

Sieger, David B. and A. B. Badiru (1993), "An Artificial Neural Network Case Study: Prediction versus Classification in a Manufacturing Application," *Computers and Industrial Engineering* 25(1–4): 381–384.

Sirinterlikci, A. (2000), Thermal Management and Prediction of Heat Checking in H-13 Die-Casting Dies, Ph.D. Dissertation, The Ohio State University.

Sirinterlikci, A. (2014), "Manufacturing Processes and Systems," in Badiru, A. B. (Ed), *Handbook of Industrial and Systems Engineering*, 2nd Edition, pp. 371–397, CRC Press/Taylor & Francis, Boca Raton, FL.

Sirinterlikci, A., Acors, C., Pogel, S., Wissinger, J., and Jimenez, M. (2012), *Antimicrobial Technologies in Design and Manufacturing of Medical Devices, SME Nanomanufacturing Conference*, Boston, MA.

Sirinterlikci, A., Karaman, A., Imamoglu, O., Buxton, G., Badger, P., and Dress, B. (2010), Role of Biomaterials in Sustainability, *2nd Annual Conference of the Sustainable Enterprises of the Future*, Pittsburgh, PA.

Timms, H.L. and Pohlen, M.F. (1970), *The Production Function in Business – Decision Systems for Production and Operations Management*, 3rd Edition, Irwin, Homewood, IL.

U.S. Patent (1986), "Apparatus for production of three-dimensional objects by stereolithography," US 4575330 A, March 11.

The DEJI Systems Model in Industry 4.0

4

SYSTEMS INTERACTIONS

Industry 4.0 is now sweeping through business and industry. It is expected that other enterprises, including government, academia, and the military, will be adopting and leveraging Industry 4.0 in the coming years. It is thus essential that product quality assurance under Industry 4.0 be fully understood. This chapter adds to the growing knowledge base of Industry 4.0 tools and techniques. The specific methodology of the chapter focuses on using the systems approach of the DEJI Systems Model in coordination with other project implementation techniques to expatiate the efficacy of Industry 4.0 as well as Quality 4.0 in diverse enterprises.

Industry 4.0 is simply the Fourth Industrial Revolution. The common catch phrase for it is Industry 4.0 (sometimes referred to as 4IR). The fundamental characteristics of Industry 4.0 is that it leverages rapid changes in technologies, the design of industries, the utilization of digital tools, technical innovation, societal preferences, and creative workforce development. Essentially, Industry 4.0 centers on the three ingredients of an organization:

- People
- Technology
- Process

The way in which these three elements interface, interact, and interconnect in a global environment is what determines the efficacy of Industry 4.0. Since many of the pieces of these elements are already available in the nooks and corners of many organizations, albeit in often-disjointed fashion, an overarching systems-based linkage is required. It is through systems thinking and implementation that the society can fully realize the benefits of Industry 4.0, which is a feature of a 21st-century operation. Although the moniker of Industry 4.0 is novel, the underlying ideas are nothing new. The tools and techniques are simply becoming more digitally based and integrated, from a systems framework.

The Fourth Industrial Revolution was officially unveiled in 2015 by a team of scientists who were charged with developing a high-tech strategy for the German government (Schwab, 2017). It has since spread widely and rapidly through various international fora (Bai et al. 2020; Philbeck and Davis, 2018), including the World Economic Forum (WEF). Quality 4.0 is a direct natural outgrowth of Industry 4.0. Using a systems framework can help bridge the talent gap in digital operations, particularly where data barriers exist.

WHAT IS QUALITY ASSURANCE?

The various definitions of quality are well understood and practiced across industry. Terms such as quality management, quality planning, and quality control are well grounded in how product quality is addressed in industry. By extension, quality assurance (QA) refers to the maintenance of a desired level of quality in a product, service, or organizational result (Agustiady and Cudney 2016). This requires paying attention to every stage of the production process, ranging from raw material integrity to the attributes of the delivered output. In this regard, a comprehensive systems approach can help identify all the elements within all the stages of the product process. In essence, quality assurance focuses on ways, techniques, and tools for preventing mistakes and defects in outputs. Quality assurance is an essential part of the broad pursuit of quality management, designed to increase the confidence that quality requirements will be fulfilled for each product. It should be noted that error detection under quality assurance precedes error detection at the quality control stage. Quality assurance focuses on quality parameters earlier in the process. For this reason, a systems tool that addresses the incipient design of the product environment is useful for preventing down-the-line errors later on in the product process. One such systems tool is the DEJI Systems Model®, which is a trademarked structure for carrying out design, evaluation, justification, and integration (Badiru, 2012, 2019).

THE TAXONOMY OF INDUSTRIAL REVOLUTION

Because Industry 4.0 represents the Fourth Industrial Revolution, it is important to touch on the preceding industrial revolutions, as summarized in Table 4.1. As can be seen in the table, industry has evolved and advanced consistently over the centuries (Badiru, 2014). This matches the characteristic commitment of the discipline of industrial engineering, which is defined by the Institute of Industrial Engineers as presented below:

Industrial engineering: The profession concerned with the design, installation, and improvement of integrated systems of people, materials, information, equipment, and energy by drawing upon specialized knowledge and skills in the mathematical, physical, and social sciences, together with the principles and methods of engineering analysis and design to specify, predict, and evaluate the results to be obtained from such systems.

DIGITIZATION IN INDUSTRY 4.0

What distinguishes Industry 4.0 from the preceding industrial revolutions is the digital environment in which modern industries operate. Of particular benefit is the emergence of artificial intelligence as the driver for improving the design and integration of modern production systems. With digital platforms, the development of smart industries is possible. With data comes more opportunities to improve operations. Industrial revolution is predicated on the availability of new radical disruptive technologies that change the course of industry. For example, the introduction of the steam engine revolutionized transportation in the 1800s. At the turn of the century, innovations in assembly lines facilitated mass-production set ups that increased a manufacturer's ability to satisfy the growing demands for various consumer products. The emergence of computers rapidly changed the work environment for the better. Each industrial revolution is ushered in by successive technological developments. Efficiency, effectiveness, and productivity improved rapidly as the society shifts from one industrial revolution to the next. In the present era of Industry 4.0, the primary drivers are centered on digital platforms. Such examples include Internet of Things (IoT),

TABLE 4.1 From the First Industrial Revolution to the Fourth Industrial Revolution

EVOLUTION OF INDUSTRY 4.0	INDUSTRIAL REVOLUTIONS			
	FIRST INDUSTRIAL REVOLUTION	SECOND INDUSTRIAL REVOLUTION	THIRD INDUSTRIAL REVOLUTION	FOURTH INDUSTRIAL REVOLUTION (INDUSTRY 4.0)
Description and Characteristics	The first industrial revolution emerged in Britain in the late 18th century. It facilitated mass production by using water flow and steam power instead of solely human and animal power. Finished goods were built with machines rather than manually.	The second industrial revolution evolved in the 19th century. It introduced assembly lines and the use of oil, natural gas, and electricity. The new sources of power sources, combined with new communications tools (e.g., telephone and telegraph) facilitated the expansion of mass production. Rudimentary automation in manufacturing processes started in this era.	The third industrial revolution started in the middle of the 20th century. It introduced the first generation of computers, advanced telecommunications, and data analysis to manufacturing processes. With computerization, the digitization of manufacturing processes started with the embedding of programmable logic controllers (PLCs) into machinery. This helped to automate machine functions. Data collection and sharing became possible and widespread.	The fourth industrial revolution, christened Industry 4.0, is characterized by increasing automation and the use of smart machines and factories. Data collection and analytics guide more efficient, effective, flexible, and productive production of products. Value chain became an item of interest. Manufacturers can more readily meet customized market demands rapidly. Mass customization emerged. Data manipulations facilitated the advancement of artificial intelligence, robotics, smart factories, and Internet of Things (IoT). Rapid and adaptive decision-making became commonplace.

cyber-physical systems, autonomous systems, robotics, data analytics, cloud computing, virtual reality, smart supply chains, and metaverse systems. Digital transformation propels organizations onto the platform of Industry 4.0. With the diverse, but cooperating digital tools, it becomes even more crucial to have a systems-based approach to integrating the elements operating under Industry 4.0. This is where the DEJI Systems Model comes into play.

PHASES OF THE DEJI SYSTEMS MODEL

With a systems viewpoint, many things are possible. In all of the foregoing discussions, product is of the essence. Product, in this sense, can be composed of physical products, service outputs, or organizational results. Therein lies the need for a systems approach. The trademarked DEJI Systems Model® presents the structured steps to achieving the intended output through an explicit integration process (Badiru, 2012, 2019). The structured steps are:

- **Design**: The design of the product, process, or organizational infrastructure. Design does not have to be limited to the conventional physical design of a product. Design can cover a variety of organizational desires and objectives, which may include both hard and soft areas of pursuit.
- **Evaluation**: The rigorous evaluation of the design with respect to the prevailing attributes of interest.
- **Justification**: The explicit assessment to justify the design for operational implementation. Not all well-designed and successfully-evaluated products are suitable and justified for real-world implementation.
- **Integration**: The rigorous commitment to align the product to the prevailing operating environment, considering the inherent limitation, preferences, and nuances of the people, tools, and processes that exist in the system.

The efficacy of the DEJI Systems Model can be affirmed in the following quote from the 2022 Proceedings of the Nigerian Academy of Engineering (NAE, 2022):

As discussed in our previous report, this systems integration approach is encapsulated by the circularity in the Design-Evaluate-Justify-Integrate (DEJI) model. The DEJI conceptualization accommodates the subtleties of the ESI stages (generation, transmission, distribution, consumption, and conservation). Circularity is a necessary condition for sustainability in systems performance as evidenced in natural ecosystems. Thus, a DEJI tariff model would be an adaptive tool for long-term sustenance of the national electricity market.

This is good for systems-themed global operations and implementations within Industry 4.0. The more systems thinking is embraced internationally, the more Systems 4.0 principles, tools, and techniques can be brought to bear on the most pressing operational challenges around the world. This is the premise of this chapter.

A missed alignment is often the reason many organizational efforts falter and fail. The essence of using the DEJI Systems Model is that it encourages and forces an organization to follow a structured path to the eventual desired outcome. The kernel of the model is captured in its operational framework presented in Figure 4.1.

Beyond the underpinning technologies, all the elements related to the actualization of Industry 4.0 are embedded in the contents of the schematic representation of the DEJI Systems Model.

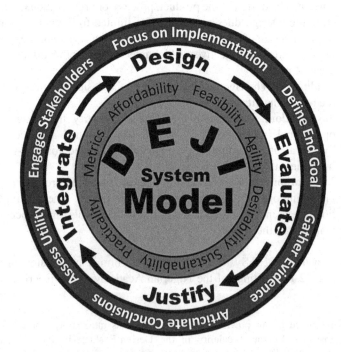

FIGURE 4.1 Schematic representation of the DEJI Systems Model.

INDUSTRY 4.0

Product Quality and Assurance

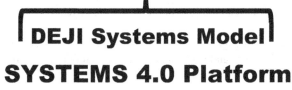

Digital Tools	Soft Tools
AI, IoT, Data Analytics, 3D Printing, Digital Twins, Machine Learning, Cloud Computing	Communication, Cooperation, Coordination, Learning Curve, Project Management

Design - Evaluation - Justification - Integration

DEJI Systems Model

SYSTEMS 4.0 Platform

FIGURE 4.2   Industry 4.0 Elements within the DEJI Systems Model.

With this systematic approach, all the existing industrial process improvement tools and techniques can be brought to bear on the outputs of Industry 4.0, in which each organization can determine and embed the prevailing and pertinent factors, attributes, and indicators. In other words, the DEJI systems model can be customized to each organization's specific needs. Examples are outlined below. Figure 4.2 shows a graphical representation of the linkage between Industry 4.0 and the DEJI Systems Model.

Design

Design Goal: Provide sustainable production processes and visible digital presence in Industry 4.0.

Metrics: Engagement with technology organizations, universities, government institutes for quality.

Expectations: Serve as a rallying point for organizational R&D; Serve as consultancy body for quality business and industry.

Evaluation

Evaluation Goal: Create a new body of technical knowledge of the nation.

Metrics: Repository of globally-accepted technologies in Industry 4.0.

Expectations: Provide quality test platforms for digital technology adoption; use engineering methodology to inform organizational strategies.

Justification

Justification Goal: Deliver R&D value to the organization, depending on desired value streams.
Metrics: Create self-sustaining quality streams.
Expectations: Increase visibility of R&D and engineering processes in engineering technology; serve as a consultancy body for industry.

Integration

Integration Goal: Align implementation to the local context of technology utilization.
Metrics: Use of computer benchmarks, research exchanges, and global partnerships.
Expectations: Engage university students, industry researchers, government experts, and integrate quality into all operations, products, and services.

Digital technologies are having a positive impact on operations. This will become even more pronounced in the future, in a post-pandemic era. Due to COVID-19 lockdowns, companies have been forced to discover the power of digital operations. There is now a remote work revolution, which can enhance the ideals of Industry 4.0. Work designs of the future must take into account how digital and conventional processes interface (Badiru and Bommer, 2017). Among the elements to pay attention to in the digital platform of Industry 4.0 are:

- Cybersecurity
- Efficacy of autonomous robots
- Augmented reality
- Big data and data analytics
- Direct digital manufacturing (aka additive manufacturing, 3D printing)
- Cloud operations
- Software designed for integration

SOFT AND HARD ELEMENTS OF INDUSTRY 4.0

A case study presented at IEOM 2021 – Mexico (IEOM 2021) reveals that over 65 percent of international companies surveyed report that soft skills are the number one impediment for implementing and sustaining Quality

4.0 under the broad pursuit of Industry 4.0. Many of the soft and hard elements to make Industry 4.0 successful and sustainable are available in the existing literature (Agustiady and Badiru, 2013; Badiru and Bommer, 2017; Badiru, 2008). Even in a digital world, human communication remains very essential. Using the Triple C Model, Badiru (2008) lays out the process of effecting communication to facilitate cooperation, which is essential for the coordination of human efforts. Without coordination, even the best-laid plans will not enmesh and the intended outputs may not be accomplished. Triple C highlights the Who, What, Why, When, Where, and How of any Industry 4.0 pursuit. Even in digitally-controlled automated systems, the humans in the loop need to be recognized vis-à-vis what the machine is expected to do when and how.

Questioning is the best approach to getting information for effective project management. Everything should be questioned. Through using upfront questions, we can preempt and avert project problems later on. Typical questions to ask under Triple C approach relate to Who, What, When, Where, Why, and How in the pursuit of communication, cooperation, and coordination to get project objectives accomplished. For the purpose of this present column, readers are referred to Badiru et al. (2019) for the nuts and bolts of implementing communication, achieving cooperation, and executing coordination across the human interfaces of a quality project. These same guidelines can be carried out for the new wave of Industry 4.0.

What really gets a project done is a sustainable commitment, which is, inherently, a human attribute in spite of whatever technological advancements exist in the project environment. Thus, soft skills are an essential component of Industry 4.0. Cooperation, the mid-point of the Triple C model, must be supported with commitment. To cooperate is to support the ideas of a project. To commit is to willingly and actively participate in project efforts repeatedly and consistently through all project obstacles. The provision of resources (budget, infrastructure, and empowerment) is one way that management can express commitment to a project.

QUALITY TOOLS FOR INDUSTRY 4.0

Standard quality-improvement tools are applicable to the environment of Industry 4.0. The most popularly-used quality tools are Six Sigma and Lean Principles. Details of both of these are provided by Agustiady and Badiru (2013) as well as Badiru and Agustiady (2021). Six Sigma is one of the several quantitative tools useful for quality assurance in Industry 4.0. Six Sigma is a process-improvement technique that seeks to eliminate or reduce variations.

The basic methodology of Six Sigma includes a five-step approach, as summarized below:

Define: Define the project environment. Initiate the project with specific reference to scope and goals. Incorporate the critical requirements of customers. In a digitally-oriented process, the step of properly defining the project and/or product becomes even more critical.

Measure: Understand the data and processes involved in the project, with a view to specifications needed for meeting customer requirements. What cannot be measured cannot be controlled. Thus, measurement is the precursor to other quality-enhancement efforts.

Analyze: Identify potential causes of problems, analyze current processes, identify relationships between inputs, processes, and outputs, and carry out data analysis. Engineering analysis is one of the fundamental components of engineering problem-solving methodology. Focus, efforts, and attention must be directed at collecting and analyzing data under Industry 4.0.

Improve: Generate solutions based on root causes and data-driven analysis while implementing effective measures. The point of doing product assessment is to improve product quality and characteristics. This becomes even more crucial under Industry 4.0 because of the fast-moving and changing digital technological developments.

Control: Righting the ship, so to speak, is the item of interest in the control phase of product management. Automation-driven digital implementations can run the risk of loss of control, should the human factor in the loop not be properly embedded. Industry 4.0 will stretch the capabilities of manufacturers. Once again, a systems approach can help identify all the nuances and requirements in the organizations in the era of Industry 4.0. Beyond Six Sigma, lean principles also have a large role to play in Industry 4.0 (Dallasega, 2021). Lean principles present guidelines and processes for streamlining operations and eliminating waste. Even though digital operations facilitate repeatable processes, the outset design of the production system must incorporate the Japanese philosophy of 5S, which is encapsulated below:

1. **Seiri** (Organize: Sort). This focuses on eliminating whatever is not needed by separating chaff from wheat, so to speak. Tools and parts not immediately needed are sorted into a secondary, but important, category. They only come up when needed. Thereby, creating room for what is immediately needed.

2. **Seiton** (Orderliness: Set in order). This focuses on organizing whatever remains neatly and arranging parts via explicit identification. Orderliness is the bastion of well-organized production systems. Organize tools for ease of access and use.

3. **Seiso** (Cleanliness: Shine). This requires that the work space be cleaned to a shining profile to facilitate visibility and accessibility.

4. **Seiketsu** (Standardize: Regiment). This recommends following a regimented template of getting things done. Sustained maintenance of production processes is expected to produce consistent outputs, whether in products, services, or results.

5. **Shitsuke** (Discipline: Sustain). This last stage of the 5S process requires that whatever gains have been achieved must be sustained with explicit discipline. This, essentially, implies making the commitment to improvement a way of normal life.

In some cases, the 5S approach is expanded to 6S, in which case safety is added as an explicit component. When Six Sigma and Lean Principles are combined, we achieve a powerful Lean-Six-Sigma practice that can enhance the overall advancement of Industry 4.0. With new technologies of artificial intelligence creeping into production systems, a combination of the DEJI systems model and Lean-Six-Sigma techniques will make the promises of Industry 4.0 realizable.

ORIGAMI IN INDUSTRY 4.0 DESIGN

Design is a foundational component of the DEJI Systems Model. Different technological tools and techniques have been brought to bear on the design process. One design technique that is rarely highlighted is Origami, the Japanese art of folding paper into decorative three-dimensional shapes and figures. As we know, design is the art of creating shapes and characteristics to meet the needs of the marketplace. Each Origami design could very well go through the stages of Design, Evaluation, Justification, and Integration. Origami is increasingly being used in engineering design requirements. With Origami, the shapes of prototypes could be explored for viability and practicality. Not only are Origami-inspired designs less expensive and faster to manufacture in two-dimensional form, but the process also opens avenues for new spheres of product scale, material properties, and mechanical movement. Many examples are already available in practice.

In space exploration, space missions need structures that are lightweight, compact, versatile, and mobile for aerial transportation. Origami-influenced space tools include antennas, photovoltaic arrays, sun shields, and solar sails. A starshade designed for exoplanet exploration is one such exciting example. A starshade would fly between a telescope and a distant star, blocking the star's light so that orbiting exoplanets could be seen and studied for signs of life. The starshade, folded to fit within a compact launch vehicle, would unfurl to a massive size once fully extended. In biomedical engineering, designers leverage Origami to make medical procedures as minimally invasive as possible. Applications include targeted drug delivery and the implanting of surgical structures deep inside a patient's body. In architecture, Origami is not only used for design aesthetics, but could also be employed to generate designs to reduce energy usage, improve structural integrity, or provide feasibility studies of alternatives. Some Origami-based designs are responsive to their environment by changing shape in reaction to light, sounds, and wind. As an example, two towers built in the United Arab Emirates in 2012 are each composed of 1,049 Origami-informed shading elements. These screens are responsive to sun exposure, opening in broad daylight to provide shade and conserve energy in and environment battered by intense heat and blowing sand. In robotic applications, compared with conventional robotics, Origami designs, when manufactured in two dimensions and then assembled into three dimensions, can be both easier to store and more cost-efficient to operate, while supporting complex computational and sensing mechanisms. For example, a vacuum-driven gripper is a soft robot that acts as an Origami skeleton that can mold around fragile items without compromising lift-or-grab power. Someday, such a technological advancement could operate on the factory floor in support of Industry 4.0 requirements. In microscopic engineering, tiny robots could be developed to injection through the tip of a needle for microsurgery operations. Origami-inspired robotic limbs could be developed to fold and unfold in response to the varying needs of a paraplegic person. In each design case and alternatives, the structured stages of the DEJI Systems Model provide design, evaluation, justification, and integration guidance for Industry 4.0, with a confirmation for the premise of Systems 4.0.

CONCLUSIONS

Variation is present in all processes, but the goal is to reduce variation as much as possible. For Six Sigma to be successful in an Industry 4.0 operation, the processes must be in control statistically and must be designed to reduce variation right from the beginning. Two types of variation are generally noticed in manufacturing operations, namely special-cause variation and common-cause variation. Special-cause variation refers to untypical events, such as production disruption. By comparison, common-cause variation is inherently embedded in all processes and can be difficult to spot and identify, particularly in a fast-changing digital environment. A root-cause analysis should be done on special-cause variation so that it can be preempted in future production runs. Management action is often needed to rectify common-cause variation. In all of these, a systems framework, using the DEJI Systems Model, can improve the efficacy of quality assurance for Industry 4.0. Through this Quality Insight column, it is anticipated that appropriate research studies will ensue to continue to promote a systems approach to quality assurance in Industry 4.0. In this context, this chapter recommends leveraging the Triple Helix of government, university, and industry collaborations. If all stakeholders cooperate and coordinate, according to the Triple C model, better outcomes can be achieved in Industry 4.0.

REFERENCES

Agustiady, Tina and A. B. Badiru (2013), *Sustainability: Utilizing Lean Six Sigma Techniques*, Taylor & Francis/CRC Press, Boca Raton, FL.

Agustiady, Tina and Elizabeth Cudney (2016), *Total Productive Maintenance: Strategies and Implementation Guide*, Taylor & Francis/CRC Press, Boca Raton, FL.

Badiru, Adedeji and Tina Agustiady (2021), *Sustainability: A Systems Engineering Approach to the Global Grand Challenge*, Taylor & Francis/CRC Press, Boca Raton, FL.

Badiru, Adedeji B. (2008), *Triple C Model of Project Management: Communication, Cooperation, and Coordination*, Taylor & Francis/CRC Press, Boca Raton, FL.

Badiru, Adedeji B. (2012), "Application of the DEJI Model for Aerospace Product Integration," *Journal of Aviation and Aerospace Perspectives (JAAP)* 2(2): 20–34, Fall.

Badiru, Adedeji B. editor (2014), *Handbook of Industrial & Systems Engineering*, 2nd Edition, Taylor & Francis/CRC Press, Boca Raton, FL.

Badiru, Adedeji B. (2019), *Systems Engineering Models: Theory, Methods, and Applications*, Taylor & Francis/CRC Press, Boca Raton, FL.

Badiru, Adedeji B., S. Abi Badiru, and I. Ade Badiru (2019), *Mechanics of Project Management: Nuts and Bolts of Project Execution*, Taylor & Francis/CRC Press, Boca Raton, FL.

Badiru, Adedeji B. and Sharon C. Bommer (2017), *Work Design: A Systematic Approach*, Taylor & Francis/CRC Press, Boca Raton, FL.

Bai, Chunguang, Patrick Dallasega, Guido Orzes, Joseph Sarkis (2020), "Industry 4.0 technologies assessment: A sustainability perspective," *International Journal of Production Economics 229*: 107776. DOI: 10.1016/j.ijpe.2020.107776. ISSN 0925-5273.

Dallasega, Patrick (2021), "The Interconnection between Lean Management and Industry 4.0 Concepts and Technologies," in *Proceedings of the 6th North American Conference on Industrial Engineering & Operations Management*, Monterrey, Mexico, November 3–5, 2021.

IEOM (2021), *Industrial and Operations Management Society, Conference Proceedings, IEOM 2021 – Mexico, 26th Industry Solutions and Industry 4.0*, Monterrey, Mexico, November 3–5, 2021.

NAE (2022), *Proceedings of the Nigerian Academy of Engineering: July 2021 – June 2022*, December 2022, Lagos, Nigeria.

Philbeck, Thomas; Nicholas Davis (2018), "The Fourth Industrial Revolution: Shaping A New Era," *Journal of International Affairs 72*(1): 17–22. ISSN 0022-197X.

Schwab, Klaus (2017), *The Fourth Industrial Revolution*, Crown Publishing Group, New York, NY. ISBN 9781524758875.

Industry 4.0 Technology Transfer Within Systems 4.0

5

INTRODUCTION

Because of its dependence on technological advancements, Industry 4.0 benefits from technology transfer partnerships. Technology can easily be misused, if it is not properly controlled. Technology evolves for beneficial purposes, but its use can be misapplied, mis-transferred, or misunderstood (Badiru et al., 2019; Badiru, 2016). Using appropriate transfer strategies can ensure that manufacturing technology avoids pitfalls that lead to failure. Technology transfer is not just about the hardware components of the technology. It can involve a combination of several components, including software (computer-based) and "skinware" (people-based). Thus, this chapter addresses the transfer of knowledge as well as the transfer of skills, specifically within the needs of Industry 4.0. When technology is transferred from a systems perspective, it becomes more encompassing to effectively dot all the T's and cross all the I's.

Why reinvent the wheel when it can be transferred and adopted from existing wheeled applications? The concepts of project management can be very helpful in planning for the adoption and implementation of new industrial technology. Due to its many interfaces, the area of industrial technology adoption and implementation is a prime candidate for the application of project planning and control techniques. Technology managers, engineers, and analysts should make an effort to take advantage of the effectiveness of project management tools. This applies the various project management techniques that have been discussed in the preceding chapters to the problem of industrial technology transfer. Project management approach is presented within the context of technology adoption and implementation for industrial development. Project management guidelines are presented for industrial technology management. The Triple C model of Communication, Cooperation, and Coordination is applied as an effective tool for ensuring the acceptance of new technology. The importance of new technologies in improving product quality and operational productivity is also discussed. The chapter also outlines the strategies for project planning and control in complex technology-based operations.

CHARACTERISTICS OF TECHNOLOGY TRANSFER

To transfer technology, we must know what constitutes technology. A working definition of technology will enable us to determine how best to transfer it. A basic question that should be asked is:

What is technology?

Technology can be defined as follows:

Technology is a combination of physical and nonphysical processes that makes use of the latest available knowledge to achieve business, service, or production goals.

Technology is a specialized body of knowledge that can be applied to achieve a mission or purpose. The knowledge concerned could be in the form of methods, processes, techniques, tools, machines, materials and procedures. Technology design, development and effective use is driven by effective utilization of human resources and effective management systems. Technological progress is the result obtained when the provision of technology is used in an effective and efficient manner to improve productivity, reduce waste, improve human satisfaction, and raise the quality of life.

Technology all by itself is useless. However, when the right technology is put to the right use, with an effective supporting management system, it can be very effective in achieving industrialization goals. Technology implementation starts with an idea and ends with a productive industrial process. Technological progress is said to have occurred when the outputs of technology, in the form of information, instrument, or knowledge that is used productively and effectively in industrial operations, lead to a lowering of costs of production, better product quality, higher levels of output (from the same amount of inputs), and higher market share. The information and knowledge involved in technological progress includes those which improve the performance of management, labor, and the total resources expended for a given activity.

Technological progress plays a vital role in improving overall national productivity. Experience in developed countries such as in the United States show that in the period 1870–1957, 90 percent of the rise in real output per man-hour can be attributed to technological progress. It is conceivable that a higher proportion of increases in per capita income is accounted for by technological change. Changes occur through improvements in the efficiency in the use of existing technology. That is, through learning and through the adaptation of other technologies, some of which may involve different collections of technological equipment. The challenge to developing countries is how to develop the infrastructure that promote, use, adapt, and advance technological knowledge.

Most of the developing nations today face serious challenges arising not only from the world-wide imbalance of dwindling revenue from industrial products and oil, but also from major changes in a world economy that is characterized by competition, imports, and exports of not only oil, but also basic technology, weapon systems, and electronics. If technology utilization is not given the right attention in all sectors of the national economy, the much-desired industrial development cannot occur or cannot be sustained. The ability of a nation to compete in the world market will, consequently, be stymied.

The important characteristics or attributes of a new technology may include productivity improvement, improved quality, cost savings, flexibility, reliability, and safety. An integrated evaluation must be performed to ensure that a proposed technology is justified both economically and technically. The scope and goals of the proposed technology must be established right from the beginning of the project. Table 5.1 summarizes some of the common "ilities" characteristics of technology transfer for a well-rounded assessment.

An assessment of a technology transfer opportunity will entail a comparison of departmental objectives with overall organizational goals in the following areas:

1. Industrial Marketing Strategy: This should identify the customers of the proposed technology. It should also address items such as market cost of proposed product, assessment of competition, and market share. Import and export considerations should be a key component of the marketing strategy.
2. Industry Growth and Long-range Expectations: This should address short-range expectations, long-range expectations, future competitiveness, future capability, and prevailing size and strength of the industry that will use the proposed technology.

TABLE 5.1 The "ilities" of Systems 4.0 Technology Transfer for Industry 4.0

CHARACTERISTICS	DEFINITIONS, QUESTIONS, AND IMPLICATIONS
Adaptability	Can the technology be adapted to fit the needs of the organization? Can the organization adapt to the requirements of the technology?
Affordability	Can the organization afford the technology in terms of first-cost, installation cost, sustainment cost, and other incidentals?
Capability	What are the capabilities of the technology with respect to what the organization needs? Can the technology meet the current and emerging needs of the organization?
Compatibility	Is the technology compatible with existing software and hardware?
Configurability	Can the technology be configured for the existing physical infrastructure available within the organization?
Dependability	Is the technology dependable enough to produce the outputs expected?
Desirability	Is the particular technology desirable for the prevailing operating environment of the organization? Are there environmental issues and/or social concerns related the technology?
Expandability	Can the technology be expanded to fit the changing needs of the organization?
Flexibility	Does the technology have flexible characteristics to accomplish alternate production requirements?
Interchangeability	Can the technology be interchanged with currently available tools and equipment in the organization? In case of operational problems, can the technology be interchanged with something else?
Maintainability	Does the organization have the wherewithal to maintain the technology?
Manageability	Does the organization have adequate management infrastructure to acquire and use the technology?
Re-configurability	When operating conditions change or organizational infrastructure change, can the technology be re-configured to meet new needs?
Reliability	Is the technology reliable in terms of technical, physical, and/or scientific characteristics?
Stability	Is the technology mature and stable enough to warrant an investment within the current operating scenario?
Sustainability	Is the organization committed enough to sustain the technology for the long haul? Is the design of the technology sound and proven to be sustainable?
Volatility	Is the technology devoid of volatile developments? Is the source of the technology devoid of political upheavals and/or social unrests?

3. National Benefit: Any prospective technology must be evaluated in terms of direct and indirect benefits to be generated by the technology. These may include product price versus value, increase in international trade, improved standard of living, cleaner environment, safer workplace, and higher productivity.

4. Economic Feasibility: An analysis of how the technology will contribute to profitability should consider past performance of the technology, incremental benefits of the new technology versus conventional technology, and value added by the new technology.

5. Capital Investment: Comprehensive economic analysis should play a significant role in the technology assessment process. This may cover an evaluation of fixed and sunk costs, cost of obsolescence, maintenance requirements, recurring costs, installation cost, space requirement cost, capital substitution options, return on investment, tax implications, cost of capital, and other concurrent projects.

6. Resource Requirements: The utilization of resources (human resources and equipment) in the pre-technology and post-technology phases of industrialization should be assessed. This may be based on material input–output flows, high value of equipment versus productivity improvement, required inputs for the technology, expected output of the technology, and utilization of technical and nontechnical personnel.

7. Technology Stability: Uncertainty is a reality in technology adoption efforts. Uncertainty will need to be assessed for the initial investment, return on investment, payback period, public reactions, environmental impact, and volatility of the technology.

8. National Productivity Improvement: An analysis of how the technology may contribute to national productivity can be verified by studying industrial throughput, efficiency of production processes, utilization of raw materials, equipment maintenance, absenteeism, learning rate, and design-to-production cycle.

INDUSTRY 4.0 AND TECHNOLOGY ADVANCEMENT

New industrial and service technologies have been gaining more attention in recent years. This is due to the high rate at which new productivity improvement technologies are being developed. The fast pace of new technologies has created difficult implementation and management problems for many organizations.

New technology can be successfully implemented only if it is viewed as a system whose various components must be evaluated within an integrated managerial framework. Such a framework is provided by a project management approach. A multitude of new technologies has emerged in recent years. It is important to consider the peculiar characteristics of a new technology before establishing adoption and implementation strategies. The justification for the adoption of a new technology is usually a combination of several factors rather than a single characteristic of the technology. The potential of a specific technology to contribute to industrial development goals must be carefully assessed. The technology assessment process should explicitly address the following questions:

> What is expected from the new technology?
> Where and when will the new technology be used?
> How is the new technology similar to or different from existing technologies?
> What is the availability of technical personnel to support the new technology?
> What administrative support is needed for the new technology?
> Who will use the new technology?
> How will the new technology be used?
> Why is the technology needed?

The development, transfer, adoption, utilization, and management of technology is a problem that is faced in one form or another by business, industry, and government establishments. Some of the specific problems in technology transfer and management include the following:

- Controlling technological change
- Integrating technology objectives
- Shortening the technology transfer time
- Identifying a suitable target for technology transfer
- Coordinating the research and implementation interface
- Formal assessment of current and proposed technologies
- Developing accurate performance measures for technology
- Determining the scope or boundary of technology transfer
- Managing the process of entering or exiting a technology
- Understanding the specific capability of a chosen technology
- Estimating the risk and capital requirements of a technology

Integrated managerial efforts should be directed at the solution of the problems stated above. A managerial revolution is needed in order to cope with the ongoing technological revolution. The revolution can be initiated by modernizing the long-standing and obsolete management culture relating to technology transfer. Among the managerial functions that will need to be addressed when developing a technology transfer strategy are the following:

1. Development of a technology transfer plan
2. Assessment of technological risk
3. Assignment/reassignment of personnel to implement the technology transfer
4. Establishment of a transfer manager and a technology transfer office. In many cases, transfer failures occur because no individual has been given the responsibility to ensure the success of technology transfer.
5. Identification and allocation of the resources required for technology transfer.
6. Setting of guidelines for technology transfer. For example,
   a. Specification of phases (Development, Testing, Transfer, etc.)
   b. Specification of requirements for inter-phase coordination
   c. Identification of training requirements
   d. Establishment and implementation of performance measurement.
7. Identify key factors (both qualitative and quantitative) associated with technology transfer and management.
8. Investigate how the factors interact and develop the hierarchy of importance for the factors.
9. Formulate a loop system model that considers the forward and backward chains of actions needed to effectively transfer and manage a given technology.
10. Track the outcome of the technology transfer.

Technological developments in many industries appear in scattered, narrow, and isolated areas within a few selected fields. This makes technology efforts to be rarely coordinated, thereby hampering the benefits of technology. The optimization of technology utilization is, thus, very difficult. To overcome this problem and establish the basis for effective technology transfer and management, an integrated approach must be followed. An integrated approach will be applicable to technology transfer between any two organizations, whether public or private.

Some nations concentrate on the acquisition of bigger, better, and faster technology. But little attention is given to how to manage and coordinate the operations of the technology once it arrives. When technology fails, it is not necessarily because the technology is deficient. Rather, it is often the communication, cooperation, and coordination functions of technology management that are deficient. Technology encompasses factors and attributes beyond mere hardware, software, and "skinware," which refers to people issues affecting the utilization of technology. This may involve social-economic and cultural issues of using certain technologies. Consequently, technology transfer involves more than the physical transfer of hardware and software. Several flaws exist in the common practices of technology transfer and management. These flaws include in the following:

- Poor fit: This relates to an inadequate assessment of the need of the organization receiving the technology. The target of the transfer may not have the capability to properly absorb the technology.
- Premature transfer of technology: This is particularly acute for emerging technologies that are prone to frequent developmental changes.
- Lack of focus: In the attempt to get a bigger share of the market or gain an early lead in the technological race, organizations frequently force technology in many incompatible directions.
- Intractable implementation problems: Once a new technology is in place, it may be difficult to locate sources of problems that have their roots in the technology transfer phase itself.
- Lack of transfer precedents: Very few precedents are available on the management of brand new technology. Managers are, thus, often unprepared for their new technology management responsibilities.
- Stuck on technology: Unworkable technologies sometimes continue to be recycled needlessly in the attempt to find the "right" usage.
- Lack of foresight: Due to the nonexistence of a technology transfer model, managers may not have a basis against which they can evaluate future expectations.
- Insensitivity to external events: Some external events that may affect the success of technology transfer may include trade barriers, taxes, and political changes.
- Improper allocation of resources: There is usually not enough resources available to allocate to technology alternatives. Thus, a technology transfer priority must be developed.

The following steps provide a specific guideline for pursuing the implementation of manufacturing technology transfer:

1. Find a suitable application.
2. Commit to an appropriate technology.
3. Perform economic justification.
4. Secure management support for the chosen technology.
5. Design the technology implementation to be compatible with existing operations.

6. Formulate project management approach to be used.
7. Prepare the receiving organization for the technology change.
8. Install the technology.
9. Maintain the technology.
10. Periodically review the performance of the technology based on prevailing goals.

TECHNOLOGY TRANSFER MODES

The transfer of technology can be achieved in various forms. Project management provides an effective means of ensuring proper transfer of technology. Three technology transfer modes are presented here to illustrate basic strategies for getting one technological product from one point (technology source) to another point (technology sink). The university–industry interaction model presented in this book can be used as an effective mechanism for facilitating technology transfer. Industrial technology application centers may be established to serve as a unified point for linking technology sources with interested targets. The center will facilitate interactions between business establishments, academic institutions, and government agencies to identify important technology needs. Technology can be transferred in one or a combination of the following strategies:

1. Transfer of complete technological products: In this case, a fully developed product is transferred from a source to a target. Very little product development effort is carried out at the receiving point. However, information about the operations of the product is fed back to the source so that necessary product enhancements can be pursued. So, the technology recipient generates product information which facilitates further improvement at the technology source. This is the easiest mode of technology transfer and the most tempting. Developing nations are particularly prone to this type of transfer. Care must be exercised to ensure that this type of technology transfer does not degenerate into "machine transfer." It should be recognized that machines alone do not constitute technology.
2. Transfer of technology procedures and guidelines: In this technology transfer mode, procedures (e.g., blueprints) and guidelines are transferred from a source to a target. The technology blueprints are implemented locally to generate the desired services and products. The use of local raw materials and manpower is encouraged for

the local production. Under this mode, the implementation of the transferred technology procedures can generate new operating procedures that can be fed back to enhance the original technology. With this symbiotic arrangement, a loop system is created whereby both the transferring and the receiving organizations derive useful benefits.

3. Transfer of technology concepts, theories, and ideas: This strategy involves the transfer of the basic concepts, theories, and ideas behind a given technology. The transferred elements can then be enhanced, modified, or customized within local constraints to generate new technological products. The local modifications and enhancements have the potential to generate an identical technology, a new related technology, or a new set of technology concepts, theories, and ideas. These derived products may then be transferred back to the original technology source as new technological enhancements. An academic institution is a good potential source for the transfer of technology concepts, theories, and ideas.

It is very important to determine the mode in which technology will be transferred for manufacturing purposes. There must be a concerted effort by people to make the transferred technology work within local infrastructure and constraints. Local innovation, patriotism, dedication, and willingness to adapt technology will be required to make technology transfer successful. It will be difficult for a nation to achieve industrial development through total dependence on transplanted technology. Local adaptation will always be necessary.

TECHNOLOGY CHANGE-OVER STRATEGIES

Any development project will require changing from one form of technology to another. The implementation of a new technology to replace an existing (or a nonexistent) technology can be approached through one of several options. Some options are more suitable than others for certain types of technologies. The most commonly used technology change-over strategies include the following:

Parallel change-over: In this case, the existing technology and the new technology operate concurrently until there is confidence that the new technology is satisfactory.

Direct change-over: In this approach, the old technology is removed totally and the new technology takes over. This method is recommended only when there is no existing technology or when both technologies cannot be kept operational due to incompatibility or cost considerations.

Phased change-over: In this incremental change-over method, modules of the new technology are gradually introduced one at a time using either direct or parallel change-over.

Pilot change-over: In this case, the new technology is fully implemented on a pilot basis in a selected department within the organization.

POST-IMPLEMENTATION EVALUATION

The new technology should be evaluated only after it has reached a steady-state performance level. This helps to avoid the bias that may be present at the transient stage due to personnel anxiety, lack of experience, or resistance to change. The system should be evaluated for the following aspects:

- Sensitivity to data errors
- Quality and productivity
- Utilization level
- Response time
- Effectiveness

TECHNOLOGY SYSTEMS INTEGRATION

With the increasing shortages of resources, more emphasis should be placed on the sharing of resources. Technology resource sharing can involve physical equipment, facilities, technical information, ideas, and related items. The integration of technologies facilitates the sharing of resources. Technology integration is a major effort in technology adoption and implementation. Technology integration is required for proper product coordination. Integration facilitates the coordination of diverse technical and managerial efforts to enhance organizational functions, reduce cost, improve productivity, and increase the utilization of resources. Technology integration ensures that all performance goals are satisfied with a minimum of expenditure of time and resources. It may require the adjustment of functions to permit sharing of resources, development of new policies to accommodate product integration, or realignment of managerial responsibilities. It can affect both hardware and software components of an organization. Important factors in technology integration include the following:

- Unique characteristics of each component in the integrated technologies
- Relative priorities of each component in the integrated technologies
- How the components complement one another
- Physical and data interfaces between the components
- Internal and external factors that may influence the integrated technologies
- How the performance of the integrated system will be measured

NATIONAL IMPERATIVE FOR TECHNOLOGY TRANSFER

The malignant policies and operating characteristics of some of the governments in underdeveloped countries have contributed to stunted growth of technology in those parts of the world. The governments in most developing countries control the industrial and public sectors of the economy. Either people work for the government or they serve as agents or contractors for the government. The few industrial firms that are privately owned depend on government contracts to survive. Consequently, the nature of the government can directly determine the nature of industrial technological progress.

The operating characteristics of most of the governments perpetuate inefficiency, corruption, and bureaucratic bungles. This has led to a decline in labor and capital productivity in the industrial sectors. Using the Pareto distribution, it can be estimated that in most government-operated companies, there are eight administrative workers for every two production workers. This creates a non-productive environment that is skewed toward hyper-bureaucracy. The government of a nation pursuing industrial development must formulate and maintain an economic stabilization policy. The objective should be to minimize the sacrifice of economic growth in the short run and while maximizing long-term economic growth. To support industrial technology transfer efforts, it is essential that a conducive national policy be developed.

More emphasis should be placed on industry diversification, training of the workforce, supporting financial structure for emerging firms and implementing policies that encourage productivity in a competitive economic environment. Appropriate foreign exchange allocation, tax exemptions, bank loans for emerging businesses, and government-guaranteed low-interest loans for potential industrial entrepreneurs are some of the favorable policies to spur growth and development of the industrial sector.

Improper trade and domestic policies have adversely affected industrialization in many countries. Excessive regulations that cause bottlenecks in industrial enterprises are not uncommon. The regulations can take the form of licensing, safety requirements, manufacturing value-added quota requirements, capital contribution by multinational firms, and high domestic production protection. Although regulations are needed for industrial operations, excessive controls lead to low returns from the industrial sectors. For example, stringent regulations on foreign exchange allocation and control have led to the closure of industrial plants in some countries. The firms that cannot acquire essential raw materials, commodities, tools, equipment, and new technology from abroad due to foreign exchange restrictions are forced to close and lay off workers.

Price controls for commodities are used very often by developing countries, especially when inflation rates for essential items are high. The disadvantages involved in price control of industrial goods include: restrictions of the free competitive power of available goods in relation to demand and supply; encouragement of inefficiency; the promotion of dual markets; the distortion of cost relationships; and an increase in administrative costs involved in producing goods and services.

CASE EXAMPLES FOR TECHNOLOGY TRANSFER

One way that a government can help facilitate industrial technology transfer involves the establishment of technology transfer centers within appropriate government agencies. A good example of this approach can be seen in the government-sponsored technology transfer program by the U.S. National Aeronautics and Space Administration (NASA). In the Space Act of 1958, the U.S. Congress charged NASA with a responsibility to provide for the widest practical and appropriate dissemination of information concerning its activities and the results achieved from those activities. With this technology transfer responsibility, technology developed in the United States' space program is available for use by the nation's business and industry.

In order to accomplish technology transfer to industry, NASA established a Technology Utilization Program (TUP) in 1962. The technology utilization program uses several avenues to disseminate information on NASA technology. The avenues include the following:

- Complete, clear, and practical documentation is required for new technology developed by NASA and its contractors. These are available to industry through several publications produced by NASA. An example is a monthly Tech Brief which outlines technology innovations. This is a source of prompt technology information for industry.
- Industrial Application Centers (IAC) were developed to serve as repositories for vast computerized data on technical knowledge. The IACs are located at academic institutions around the country. All the centers have access to a large data base containing millions of NASA documents. With this data base, industry can have access to the latest technological information quickly. The funding for the centers are obtained through joint contributions from several sources, including NASA, the sponsoring institutions, and state government subsidies. Thus, the centers can provide their services at very reasonable rates.
- NASA operates a Computer Software Management and Information Center (COSMIC) to disseminate computer programs developed through NASA projects. COSMIC, which is located at a university, has a library of thousands of computer programs. The center publishes an annual index of available software.

In addition to the specific mechanisms discussed above, NASA undertakes Application Engineering Projects. Through these projects, NASA collaborates with industry to modify aerospace technology for use in industrial applications. To manage the application projects, NASA established a Technology Application Team (TAT), consisting of scientists and engineers from several disciplines. The team interacts with NASA field centers, industry, universities, and government agencies. The major mission of the team interactions is to define important technology needs and identify possible solutions within NASA. NASA applications engineering projects are usually developed in a five-phase approach with go or no-go decisions made by NASA and industry at the completion of each phase. The five phases are outlined below:

1. NASA and the Technology Applications Team meet with industry associations, manufacturers, university researchers, and public sector agencies to identify important technology problems that might be solved by aerospace technology.

2. After a problem is selected, it is documented and distributed to the Technology Utilization Officer at each of NASA's field centers. The officer in turn distributes the description of the problem to the appropriate scientists and engineers at the center. Potential solutions are forwarded to the team for review. The solutions are then screened by the problem originator to assess the chances for technical and commercial success.
3. The development of partnerships and a project plan to pursue the implementation of the proposed solution. NASA joins forces with private companies and other organizations to develop an applications engineering project. Industry participation is encouraged through a variety of mechanisms such as simple letters of agreement or joint endeavor contracts. The financial and technical responsibilities of each organization are specified and agreed upon.
4. At this point, NASA's primary role is to provide technical assistance to facilitate utilization of the technology. The costs for these projects are usually shared by NASA and the participating companies. The proprietary information provided by the companies and their rights to new discoveries are protected by NASA.
5. The final phase involves the commercialization of the product. With the success of commercialization, the project would have widespread impact. Usually, the final product development, field testing, and marketing are managed by private companies without further involvement from NASA.

Through this well-coordinated government-sponsored technology transfer program, NASA has made significant contributions to the U.S. industry. The results of NASA's technology transfer abound in numerous consumer products, either in subtle forms or in clearly identifiable forms. Food preservation techniques constitute one area of NASA's technology transfer that has had a significant positive impact on the society. Although the specific organization and operation of the NASA technology transfer programs have changed in name or in deed over the years, the basic descriptions outlined above remain a viable template for how to facilitate manufacturing technology transfer. Other nations can learn from NASA's technology transfer approach. In a similar government-backed strategy, the US Air Force Research Lab (AFRL) also has very structured programs for transferring non-classified technology to the industrial sector.

The major problem in developing nations is not the lack of good examples to follow. Rather, the problem involves not being able to successfully manage and sustain a program that has proven successful in other nations. It is believed that a project management approach can help in facilitating success with manufacturing technology transfer efforts.

PATHWAY FOR INDUSTRIAL ADVANCEMENT

Most of the developing nations depend on technologies transferred from developed nations to support their industrial base. This is partly due to a lack of local research and development programs, development funds, and workforce needed to support such activity. Advanced technology is desired by most industries in developing countries because of its potential to increase output. The adaptability of advanced technology to industries in a developing country is a complex and difficult task. Evidence in most manufacturing firms that operate in developing countries reveal that advanced technology can lead to machine down time because the local plants do not have the necessary maintenance and repair facilities to support the use of advanced technology.

In some situations, most firms cannot afford the high cost of maintenance associated with the use of foreign technology. One way to solve the transfer of technology problem is by establishing local design centers for developing nation's industrial sectors; such centers will design and adapt technology for local usage. In addition, such centers will also work on adapting full assembled machinery from developed countries. However, the fertile ground for the introduction of appropriate technology is where people are already organized under a good system of government, production, marketing, and continuing improvement in the standard of living. Developing countries must place more emphasis on the production of useful, consumable goods and services. One useful strategy to ensure a successful transfer of technology is by providing training services that will ensure proper repair and maintenance of technology hardware. It is important that a nation trying to transfer technology should have access to a broad-based body of technical information and experience. A plan of technical information sharing between suppliers and users must be assured. The transfer of technology also requires a reliable liaison between the people who develop the ideas, their agents, and the people who originate the concepts. Technology transfer is only complete when the technology becomes generally accepted in the workplace. Local efforts are needed in tailoring technological solutions to local problems. Technicians and engineers must be trained to assume the role of technology custodians so that implementation and maintenance problems are minimized. A strategy for minimizing the technology transfer disconnection is to set up central repair shops dedicated to making spare parts and repairing equipment on a timely basis to reduce industrial machine down time. If the utilization level of equipment is increased, there will be an increase in the productive capacity of the manufacturer. Improving maintenance and repair centers in developing countries will provide an effective way of assisting emerging firms in developing countries where dependence on transferred technology is prevalent. There should also be a strategy to develop appropriate local technology to support the goals of industrialization. This is important because fully transferred technology may not be fully suitable or compatible with local product specifications. For example, many nations have experienced the failure of transferred food-processing technology because the technology was not responsive to the local diets, ingredients, and food-preparation practices. One way to accomplish the development of local technology is to encourage joint research efforts between academic institutions and industrial firms. Some of the chapters in this book explicitly address university–industry collaborations. The design centers suggested earlier can help in this process. In addition to developing new local technologies, existing technologies should be calibrated for local usage and the higher production level required for industrialization.

The government of developing nations must assume leadership roles in encouraging research and development activities, awarding research grants to universities and private organizations geared toward seeking better ways for developing and adapting technologies for local usage. Effective innovations and productivity improvement cannot happen without adequate public and private sector policies. A nation that does not have an effective policy for productivity management and technology advancement will always find itself in a cycle of unstable economy and business crisis. Increases in real product capital, income level, and quality of life are desirable goals that are achievable through effective policies that are executed properly. The following recommendations are offered to encourage industrial growth and technological progress:

1. Encourage a free enterprise system that believes in and practice fair competition. Discourage protectionism and remove barriers to allow free trade.
2. Avoid nationalization of assets of companies jointly developed by citizens of developing countries and multinationals. Encourage joint industrial ventures among nations.
3. Both public and private sectors of the economy should encourage and invest in improving national education standards for citizen at various levels.
4. Refrain from dependence on borrowed money and subsidy programs. Create productive enterprises locally that provide essential commodities for local consumptions and exports.
5. Both public and private sectors should invest more on systems and programs, research and development that generate new breakthrough in technology and methods for producing food rather than war instruments.

6. The public sector should establish science and technology centers to foster the development of new local technology, productivity management techniques, and production methodologies.

7. Encourage strong partnership between government, industry and academic communities in formulating and executing national development programs.

8. Governments and financial institutions should provide low-interest loans to entrepreneurs willing to take risk in producing essential goods and services through small scale industries.

9. Implement a tax structure that is equitable and one that provides incentives for individuals and businesses that are working to expand employment opportunities, and increase the final output of the national economy.

10. Refrain from government control of productive enterprises. Such controls only create grounds for fraud and corruption. Excessive regulations should be discouraged.

11. Periodically assess the ratio of administrative workers to production workers, administrative workers to service workers, in both private and public sectors. Implement actions to reduce excessive administrative procedures and bureaucratic bottlenecks that impede productivity and technological progress.

12. Encourage organizations and firms to develop and implement strategies, methods, and techniques in a framework of competitive and long-term performance.

13. Trade policy laws and regulations should be developed and enforced in a framework that recognizes fair competition in a global economy.

14. Create a national productivity, science and technology council to facilitate the implementation of good programs, enhance cooperation between private and public sectors of the economy, redirect the economy toward growth strategies and encourage education and training of the workforce.

15. Implement actions that insure stable fiscal, monetary, and income policies. Refrain from wage and price control by political means. Let the elements of the free enterprise system control the inflation rate, wages and income distribution.

16. Encourage moral standards that take pride in excellence, work ethics, and a value system that encourages pride in consumer products produced locally.

17. Encourage individuals and business to protect full employment programs, maintain income levels by investing in local ventures rather than exporting capital abroad.

18. Both the public and private sectors of the economy should encourage and invest in re-training of the workforce as new technology and techniques are introduced for productive activities.
19. Make use of the expertise of nations that are professionally based abroad. This is an excellent source of expertise for local technology development.
20. Arrange for annual conferences, seminars, and workshops to exchange ideas between researchers, entrepreneurs, practitioners, and managers with the focus on the processes required for industrial development.

PICK CHART FOR TECHNOLOGY SELECTION

The question of which technology is appropriate to transfer in or transfer out is relevant for technology transfer considerations. While several methods of technology selection are available, this book recommends methods that combine qualitative and quantitative factors. The Analytical Hierarchy Process (AHP) is one such method. Another useful, albeit one that is less publicized, is the PICK chart. The PICK chart was originally developed by Lockheed Martin to identify and prioritize improvement opportunities in the company's process improvement applications. The technique is just one of the several decision tools available in process improvement endeavors. It is a very effective technology selection tool used to categorize ideas and opportunities. The purpose is to qualitatively help identify the most useful ideas. A 2×2 grid is normally drawn on a white board or large flip-chart. Ideas that were written on sticky notes by team members are placed on the grid based on a group assessment of the payoff relative the level of difficulty. The PICK acronym comes from the labels for each of the quadrants of the grid: Possible (easy, low payoff), Implement (easy, high payoff), Challenge (hard, high payoff), and Kill (hard, low payoff). The PICK chart quadrants are summarized as follows:

Possible (easy, low payoff): Third quadrant
Implement (easy, high payoff): Second quadrant
Challenge (hard, high payoff): First quadrant
Kill (hard, low payoff): Fourth quadrant

The primary purpose is to help identify the most useful ideas, especially those that can be accomplished immediately with little difficulty. These are called

"Just-Do-Its." The PICK process is normally done subjectively by a team of decision-makers under a group decision process. This can lead to bias and pro-tracted debate of where each item belongs. It is desired to improve the efficacy of the process by introducing some quantitative analysis. Badiru and Thomas (2013) present a methodology to achieve a quantification of the PICK selec-tion process. The PICK chart is often criticized for its subjective rankings and lack of quantitative analysis. The approach presented by Badiru and Thomas (2013) alleviates such concerns by normalizing and quantifying the process of integrating the subjective rankings by those involved in the group PICK process. Human decision is inherently subjective. All we can do is to develop techniques to mollify the subjective inputs rather than compounding them with subjective summarization.

QUANTIFICATION METHODOLOGY

The placement of items into one of the four categories in a PICK chart is done through expert ratings, which are often subjective and non-quantitative. In order to apply some quantitative basis to the PICK chart analysis, Badiru and Thomas (2013) present the methodology of dual numeric scaling on the impact and dif-ficulty axes. Suppose each technology is ranked on a scale of one to ten and plotted accordingly on the PICK chart. Then, each project can be evaluated on a binomial pairing of the respective rating on each scale. Note that a high rating along the x axis is desirable while a high rating along the y axis is not desirable. Thus, a composite rating involving x and y must account for the adverse effect of high values of y. A simple approach is to define $y' = (11-y)$, which is then used in the composite evaluation. If there are more factors involved in the overall pro-ject selection scenario, the other factors can take on their own lettered labeling (e.g., a, b, c, z, etc.). Then, each project will have an n-factor assessment vector. In its simplest form, this approach will generate a rating such as the following:

$$PICK_{R,i}\left(x, y'\right) = x + y'$$

where $PICK_{R,i}(x,y)$ = PICK rating of project i (i = 1, 2, 3, …., n); n = number of project under consideration; x = rating along the impact axis ($1 \leq x \leq 10$); y = rating along the difficulty axis $1 \leq y \leq 10$); $y' = (11-y)$

If $x + y'$ is the evaluative basis, then each technology's composite rating will range from 2 to 20, 2 being the minimum and 20 being the maximum pos-sible. If $(x)(y)$ is the evaluative basis, then each project's composite rating will range from 1 to 100. In general, any desired functional form may be adopted for the composite evaluation. Another possible functional form is:

$$PICK_{R,i}(x, y'') = f(x, y'')$$
$$= (x + y'')^2,$$

where y'' is defined as needed to account for the converse impact of the axes of difficulty. The above methodology provides a quantitative measure for translating the entries in a conventional PICK chart into an analytical technique to rank the technology alternatives, thereby reducing the level of subjectivity in the final decision. The methodology can be extended to cover cases where a technology has the potential to create negative impacts, which may impede organizational advancement.

The quantification approach facilitates a more rigorous analytical technique compared to traditional subjective approaches. One concern is that although quantifying the placement of alternatives on the PICK chart may improve the granularity of relative locations on the chart, it still does not eliminate the subjectivity of how the alternatives are assigned to quadrants in the first place. This is a recognized feature of many decision tools. This can be mitigated by the use of additional techniques that aid decision-makers to refine their choices. The analytic hierarchy process (AHP) could be useful for this purpose. Quantifying subjectivity is a continuing challenge in decision analysis. The PICK chart quantification methodology offers an improvement over the conventional approach.

Although the PICK chart has been used extensively in industry, there are few published examples in the open literature. The quantification approach presented by Badiru and Thomas (2013) may expand interest in and applications of the PICK chart among technology researchers and practitioners. The steps for implementing a PICK Chart are summarized below:

Step 1: On a chart, place the subject question. The question needs to be asked and answered by the team at different stages to be sure that the data that is collected is relevant.

Step 2: Put each component of the data on a different note (like a post-it) or small cards. These notes should be arranged on the left side of the chart.

Step 3: Each team member must read all notes individually and consider its importance. The team member should decide whether the element should or should not remain a fraction of the significant sample. The notes are then removed and moved to the other side of the chart. Now, the data is condensed enough to be processed for a particular purpose by means of tools that allow groups to reach a consensus on priorities of subjective and qualitative data.

Step 4:   Apply the quantification methodology presented above to normalize the qualitative inputs of the team.

TECHNOLOGY INTEGRATION FOR INDUSTRY 4.0

Technology is at the intersection of efficiency, effectiveness, and productivity. Efficiency provides the framework for quality in terms of resources and inputs required to achieve the desired level of quality. Effectiveness comes into play with respect to the application of product quality to meet specific needs and requirements of an organization. Productivity is an essential factor in the pursuit of quality as it relates to the throughput of a production system. To achieve the desired levels of quality, efficiency, effectiveness, and productivity, a new technology integration framework must be adopted. This section presents a technology integration model for design, evaluation, justification, and integration (DEJI) based on the product development application presented by Badiru (2012). The model is relevant for research and development efforts in industrial development and technology applications. The DEJI (Design, Evaluation, Justification, and Integration) model encourages the practice of building quality into a product right from the beginning so that the product or technology integration stage can be more successful.

DESIGN FOR TECHNOLOGY IMPLEMENTATION

The design of quality in product development should be structured to follow point-to-point transformations. A good technique to accomplish this is the use of state-space transformation, with which we can track the evolution of a product from the concept stage to a final product stage. For the purpose of product quality design, the following definitions are applicable:

Product state: A state is a set of conditions that describe the product at a specified point in time. The *state* of a product refers to a performance characteristic of the product which relates input to output such that a knowledge of the input function over time and the state of the product at time $t = t_0$ determines the expected output for $t \geq t_0$. This is particularly important for assessing where the product stands in the context of new technological developments and the prevailing operating environment.

Product state-space: A product *state-space* is the set of all possible states of the product lifecycle. State-space representation can solve product design problems by moving from an initial state to another state, and eventually to the desired end-goal state. The movement from state to state is achieved by means of actions. A goal is a description of an intended state that has not yet been achieved. The process of solving a product problem involves finding a sequence of actions that represents a solution path from the initial state to the goal state. A state-space model consists of state variables that describe the prevailing condition of the product. The state variables are related to inputs by mathematical relationships. Examples of potential product state variables include schedule, output quality, cost, due date, resource, resource utilization, operational efficiency, productivity throughput, and technology alignment. For a product described by a system of components, the state-space representation can follow the quantitative metric below:

$$Z = f\left(z, \ x\right); \quad Y = g\left(z, \ x\right)$$

where f and g are vector-valued functions. The variable Y is the output vector while the variable x denotes the inputs. The state vector Z is an intermediate vector relating x to y. In generic terms, a product is transformed from one state to another by a driving function that produces a transitional relationship given by:

$$S_s = f(x \mid S_p) + e,$$

where S_s = subsequent state; x = state variable; S_p = the preceding state; e = error component.

The function f is composed of a given action (or a set of actions) applied to the product. Each intermediate state may represent a significant milestone in the project. Thus, a descriptive state-space model facilitates an analysis of what actions to apply in order to achieve the next desired product state. The state-space representation can be expanded to cover several components within the technology integration framework. Hierarchical linking of product elements provides an expanded transformation structure. The product state can be expanded in accordance with implicit requirements. These requirements might include grouping of design elements, linking precedence requirements (both technical and procedural), adapting to new technology developments, following required communication links, and accomplishing reporting requirements. The actions to be taken at each state depend on the prevailing product conditions. The nature of subsequent alternate states depends on what actions are implemented. Sometimes there are multiple paths that can lead to the desired end result. At other times, there exists only one unique path to the desired objective. In conventional practice, the characteristics of the future states can only be recognized after the fact, thus making it impossible to develop adaptive plans. In the implementation of the **DEJI** model, adaptive plans can be achieved because the events occurring within and outside the product state boundaries can be taken into account. If we describe a product by P state variables s_i, then the composite state of the product at any given time can be represented by a vector $\mathbf{S}$ containing P elements. That is,

$$\mathbf{S} = \{s_1, s_2, \ldots, s_P\}$$

The components of the state vector could represent either quantitative or qualitative variables (e.g., cost, energy, color, time). We can visualize every state vector as a point in the state-space of the product. The representation is unique since every state vector corresponds to one and only one point in the state-space. Suppose we have a set of actions (transformation agents) that we can apply to the product information so as to change it from one state to another within the project state-space. The transformation will change a state vector into another state vector. A transformation may be a change in raw material or a change in design approach. The number of transformations available for a product characteristic may be finite or unlimited. We can construct trajectories that describe the potential states of a product evolution as we apply successive transformations with respect to technology forecasts. Each transformation may be repeated as many times as needed. Given an initial state $\mathbf{S}_0$, the sequence of state vectors is represented by the following:

$$\mathbf{S}_n = T_n(\mathbf{S}_{n-1}).$$

The state-by-state transformations are then represented as $\mathbf{S}_1 = T_1(\mathbf{S}_0)$; $\mathbf{S}_2 = T_2(\mathbf{S}_1)$; $\mathbf{S}_3 = T_3(\mathbf{S}_2)$; ... .; $\mathbf{S}_n = T_n(\mathbf{S}_{n-1})$. The final State, $\mathbf{S}_n$, depends on the initial state $\mathbf{S}$ and the effects of the actions applied.

EVALUATION OF TECHNOLOGY

A product can be evaluated on the basis of cost, quality, schedule, and meeting requirements. There are many quantitative metrics that can be used in evaluating a product at this stage. Learning curve productivity is one relevant technique that can be used because it offers an evaluation basis of a product with respect to the concept of growth and decay. The half-life extension (Badiru, 2012) of the basic learning is directly applicable because the half-life of the technologies going into a product can be considered. In today's technology-based operations, retention of learning may be threatened by fast-paced shifts in operating requirements. Thus, it is of interest to evaluate the half-life properties of new technologies as they impact the overall product quality. Information about the half-life can tell us something about the sustainability of learning-induced technology performance. This is particularly useful for designing products whose life cycles stretch into the future in a high-tech environment.

JUSTIFICATION OF TECHNOLOGY

We need to justify a program on the basis of quantitative value assessment. The Systems Value Model (SVM) is a good quantitative technique that can be used here for project justification on the basis of value. The model provides a heuristic decision aid for comparing project alternatives. It is presented here again for the present context. Value is represented as a deterministic vector function that indicates the value of tangible and intangible attributes that characterize the project. It is represented as $V = f(A_1, A_2, ..., A_p)$, where V is the assessed value and the A values are quantitative measures or attributes. Examples of product attributes are quality, throughput, manufacturability, capability, modularity, reliability, interchangeability, efficiency, and cost performance. Attributes are considered to be a combined function of factors. Examples of product factors are market share, flexibility, user acceptance, capacity utilization, safety, and design functionality. Factors are themselves considered to be composed of indicators. Examples of indicators are debt ratio, acquisition volume, product responsiveness, substitutability, lead time, learning curve, and scrap volume. By combining the above definitions, a composite measure of the operational value of a product can be quantitatively assessed. In addition to the quantifiable factors, attributes, and indicators that impinge upon overall project value, the human-based subtle factors should also be included in assessing overall project value.

INTEGRATION OF TECHNOLOGY

Without being integrated, a system will be in isolation and it may be worthless. We must integrate all the elements of a system on the basis of alignment of functional goals. The overlap of systems for integration purposes can conceptually be viewed as projection integrals by considering areas bounded by the common elements of sub-systems. Quantitative metrics can be applied at this stage for effective assessment of the technology state. Trade-off analysis is essential in technology integration. Pertinent questions include the following:

What level of trade-offs on the level of technology are tolerable?
What is the incremental cost of more technology?
What is the marginal value of more technology?
What is the adverse impact of a decrease in technology utilization?

What is the integration of technology over time? In this respect, an integral of the form below may be suitable for further research:

$$I = \int_{t_1}^{t_2} f(q)\,dq,$$

where I = integrated value of quality, $f(q)$ = functional definition of quality, t_1 = initial time, and t_2 = final time within the planning horizon.

Presented below are guidelines and important questions relevant for technology integration:

- What are the unique characteristics of each component in the integrated system?
- How do the characteristics complement one another?
- What physical interfaces exist among the components?
- What data/information interfaces exist among the components?
- What ideological differences exist among the components?
- What are the data flow requirements for the components?

- What internal and external factors are expected to influence the integrated system?
- What are the relative priorities assigned to each component of the integrated system?
- What are the strengths and weaknesses of the integrated system?
- What resources are needed to keep the integrated system operating satisfactorily?
- Which organizational unit has primary responsibility for the integrated system?

The recommended approach of the DEJI model will facilitate a better alignment of product technology with future development and needs. The stages of the model require research for each new product with respect to design, evaluation, justification, and integration. Existing analytical tools and techniques can be used at each stage of the model.

CONCLUSIONS

Technology transfer is a great avenue to advancing industrialization. This chapter has presented a variety of principles, tools, techniques, and strategies useful for managing technology transfer. Of particular emphasis in the chapter is the management aspects of technology transfer. The technical characteristics of the technology of interest are often well understood. What is often lacking is an appreciation of the technology management requirements for achieving a successful technology transfer. This chapter presents the management aspects of manufacturing technology transfer.

REFERENCES

Badiru, Adedeji B. (2012), "Application of the DEJI Model for Aerospace Product Integration," *Journal of Aviation and Aerospace Perspectives (JAAP)* 2(2): 20–34, Fall 2012.

Badiru, Adedeji B. (2016), *Global Manufacturing Technology Transfer: Africa-USA Strategies, Adaptations, and Management*, Taylor & Francis/CRC Press, Boca Raton, FL.

Badiru, Adedeji B., Oye Ibidapo-Obe, and Babs J. Ayeni (2019), *Manufacturing and Enterprise: An Integrated Systems Approach*, Taylor & Francis/CRC Press, Boca Raton, FL.

Badiru, Adedeji B. and Marlin Thomas (2013), "Quantification of the PICK Chart for Process Improvement Decisions," *Journal of Enterprise Transformation 3*(1): 1–15.

Technology Cost Analysis in Systems 4.0

6

TECHNOLOGY COST DEFINITIONS

This chapter covers economic aspects of Industry 4.0 projects from a Systems 4.0 perspectives. Basic cash flow analysis is presented. Other topics covered include cost concepts, cost estimation, cost monitoring, budgeting allocation, and inflation. Of particular interest in the chapter is the equity break-even point formula derived by Badiru (2016). Contemporary technology is the basis for Industry 4.0. As a result, from the Systems 4.0 perspective, we must understand the cost and value of technology, particularly over time. Cost escalation, value appreciation, and the decline of effectiveness are of interest in Systems 4.0 for the purpose of advancing Industry 4.0. Cost management in a technology environment refers to the functions required to maintain effective financial control of the technology throughout its life cycle. The conventional techniques of engineering economic analysis are adapted in this chapter for the purpose of technology cost analysis. Cost is the bottom line in every organization endeavor. Thus, a robust cost analysis methodology is a key requirement in the Systems 4.0 approach to Industry 4.0. There are several cost concepts that influence the economic aspects of managing Industry 4.0 technology projects. Within a given scope of analysis, there may be a combination of different types of cost aspects to consider. These cost aspects include the ones explained below:

> **Actual Cost of Work Performed**. This represents the cost actually incurred and recorded in accomplishing the work performed within a given time period.

DOI: 10.1201/9781003312277-6

Applied Direct Cost. This represents the amounts recognized in the time period associated with the consumption of labor, material, and other direct resources, without regard to the date of commitment or the date of payment. These amounts are to be charged to work-in-process (WIP) when resources are actually consumed, material resources are withdrawn from inventory for use, or material resources are received and scheduled for use within 60 days.

Budgeted Cost for Work Performed. This is the sum of the budgets for completed work plus the appropriate portion of the budgets for level of effort and apportioned effort. Apportioned effort is effort that by itself is not readily divisible into short-span work packages but is related in direct proportion to measured effort.

Budgeted Cost for Work Scheduled. This is the sum of budgets for all work packages and planning packages scheduled to be accomplished (including work in process) plus the amount of level of effort and apportioned effort scheduled to be accomplished within a given period of time.

Direct Cost. This is a cost that is directly associated with actual operations of a project. Typical sources of direct costs are direct material costs and direct labor costs. Direct costs are those that can be reasonably measured and allocated to a specific component of a project.

Economies of Scale. This refers to a reduction of the relative weight of the fixed cost in total cost by increasing output quantity. This helps to reduce the final unit cost of a product. Economies of scale is often simply referred to as the savings due to *mass production*.

Estimated Cost at Completion. This is the actual direct cost, plus indirect costs that can be allocated to the contract, plus the estimate of costs (direct and indirect) for authorized work remaining.

First Cost. This is the total initial investment required to initiate a project or the total initial cost of the equipment needed to start the project.

Fixed Cost. This is a cost incurred irrespective of the level of operation of a project. Fixed costs do not vary in proportion to the quantity of output. Example of costs that make up the fixed cost of a project are administrative expenses, certain types of taxes, insurance cost, depreciation cost, and debt servicing cost. These costs usually do not vary in proportion to quantity of output.

Incremental Cost. This refers to the additional cost of changing the production output from one level to another. Incremental costs are normally variable costs.

Indirect Cost. This is a cost that is indirectly associated with project operations. Indirect costs are those that are difficult to assign to specific components of a project. An example of an indirect cost is the cost of computer hardware and software needed to manage project operations. Indirect costs are usually calculated as a percentage of a component of direct costs. For example, the direct costs in an organization may be computed as 10 percent of direct labor costs.

Life-Cycle Cost. This is the sum of all costs, recurring and nonrecurring, associated with a project during its entire life cycle.

Maintenance Cost. This is a cost that occurs intermittently or periodically for the purpose of keeping project equipment in good operating condition.

Marginal Cost. This is the additional cost of increasing production output by one additional unit. The marginal cost is equal to the slope of the total cost curve or line at the current operating level.

Operating Cost. This is a recurring cost needed to keep a project in operation during its life cycle. Operating costs may consist of such items as labor cost, material cost, and energy cost.

Opportunity Cost. This is the cost of forgoing the opportunity to invest in a venture that would have produced an economic advantage. Opportunity costs are usually incurred due to limited resources that make it impossible to take advantage of all investment opportunities. It is often defined as the cost of the best rejected opportunity. Opportunity costs can be incurred due to a missed opportunity rather than due to an intentional rejection. In many cases, opportunity costs are hidden or implied because they typically relate to future events that cannot be accurately predicted.

Overhead Cost. This is a cost incurred for activities performed in support of the operations of a project. The activities that generate overhead costs support the project efforts rather than contribute directly to the project goal. The handling of overhead costs varies widely from company to company. Typical overhead items are electric power cost, insurance premiums, cost of security, and inventory carrying cost.

Standard Cost. This is a cost that represents the normal or expected cost of a unit of the output of an operation. Standard costs are established in advance. They are developed as a composite of several component costs such as direct labor cost per unit, material cost per unit, and allowable overhead charge per unit.

Sunk Cost. This is a cost that occurred in the past and cannot be recovered under the present analysis. Sunk costs should have no bearing on the prevailing economic analysis and project decisions. Ignoring sunk costs is always a difficult task for analysts. For example, if $950,000 was spent four years ago to buy a piece of equipment for a technology-based project, a decision on whether or not to replace the equipment now should not consider that initial cost. But uncompromising analysts might find it difficult to ignore that much money. Similarly, an individual making a decision on selling a personal automobile would typically try to relate the asking price to what was paid for the automobile when it was acquired. This is wrong under the strict concept of sunk costs.

Total Cost. This is the sum of all the variable and fixed costs associated with a project.

Variable Cost. This is a cost that varies in direct proportion to the level of operation or quantity of output. For example, the costs of material and labor required to make an item will be classified as variable costs because they vary with changes in level of output.

COST AND CASH FLOW ANALYSIS

The basic reason for performing economic analysis is to make a choice between mutually exclusive projects that are competing for limited resources. The cost performance of each project will depend on the timing and levels of its expenditures. The techniques of computing cash flow equivalence permit us to bring competing project cash flows to a common basis for comparison. The common basis depends on the prevailing interest rate. Two cash flows that are equivalent at a given interest rate will not be equivalent at a different interest rate. The basic techniques for converting cash flows from one point in time to another are presented in the next section. Cash flow conversion involves the transfer of project funds from one point in time to another. The following basic variables are typically included in the cash flow analysis and conversion methodologies.

- Interest rate per period
- Number of interest periods
- Present sum of money
- Future sum of money
- Uniform end-of-period cash receipt or disbursement
- A uniform arithmetic gradient increase in period-by-period payments or disbursements

In many cases, the interest rate used in performing economic analysis is set equal to the minimum attractive rate of return (MARR) of the decision-maker.

The MARR is also sometimes referred to as *hurdle rate, required internal rate of return* (IRR), *return on investment* (ROI), or *discount rate*. The value of MARR is chosen with the objective of maximizing the economic performance of a project. Since this is focused book, the computational details of engineering economic analysis are not presented in this chapter. Readers are referred to the several and extensive literature available on the pertinent topics. Recommended references include Badiru (1996, 2009, 2016, 2019) and Badiru and Omitaomu (2007, 2011).

TECHNOLOGY BREAK-EVEN ANALYSIS

Break-even analysis refers to the determination of the balanced performance level where project income is equal to project expenditure. The total cost of an operation is expressed as the sum of the fixed and variable costs with respect to output quantity. That is,

$$TC(x) = FC + VC(x)$$

where x is the number of units produced, $TC(x)$ is the total cost of producing x units, FC is the total fixed cost, and $VC(x)$ is the total variable cost associated with producing x units. The total revenue resulting from the sale of x units is defined as

$$TR(x) = px$$

where p is the price per unit. The profit due to the production and sale of x units of the product is calculated as

$$P(x) = TR(x) - TC(x)$$

The break-even point of an operation is defined as the value of a given parameter that will result in neither profit nor loss. The parameter of interest may be the number of units produced, the number of hours of operation, the number of units of a resource type allocated, or any other measure of interest. At the break-even point, we have the following relationship:

$$TR(x) = TC(x) \text{ or } P(x) = 0$$

In some cases, there may be a known mathematical relationship between cost and the parameter of interest. For example, there may be a linear cost relationship between the total cost of a project and the number of units produced. The cost expressions facilitate straightforward break-even analysis. When two project alternatives are compared, the break-even point refers to the point of indifference between the two alternatives.

PROFIT RATIO ANALYSIS

Break-even charts offer opportunities for several different types of analysis. In addition to the break-even points, other measures of worth or criterion measures may be derived from the charts. A measure, called *profit ratio* (Badiru, 2016), is presented here for the purpose of obtaining a further comparative basis for competing projects. Profit ratio is defined as the ratio of the profit area to the sum of the profit and loss areas in a break-even chart. That is,

$$\text{Profit ratio} = \frac{\text{Area of profit region}}{\text{Area of profit region} + \text{Area of loss region}}$$

The profit ratio may be used as a criterion for selecting among project alternatives. If this is done, the profit ratios for all the alternatives must be calculated over the same values of the independent variable. The project with the highest profit ratio will be selected as the desired project.

The profit ratio approach evaluates the performance of each alternative over a specified range of operating levels. Most of the existing evaluation methods use single-point analysis with the assumption that the operating condition is fixed at a given production level. The profit ratio measure allows an analyst to evaluate the net yield of an alternative given that the production level may shift from one level to another. An alternative, for example, may operate at a loss for most of its early life, while it may generate large incomes to offset the losses in its later stages. Conventional methods cannot easily capture this type of transition from one performance level to another. In addition to being used to compare alternate projects, the profit ratio may also be used for evaluating the economic feasibility of a single project. In such a case, a decision rule may be developed. An example of such a decision rule is:

If profit ratio is greater than 75 percent, accept the project.
If profit ratio is less than or equal to 75 percent, reject the project.

TECHNOLOGY INVESTMENT ANALYSIS

Many capital investment projects are financed with external funds. A careful analysis must be conducted to ensure that the amortization schedule can be handled by the organization involved. A computer program such as GAMPS (graphic evaluation of amortization payments) might be used for this purpose (Badiru, 1988). The program analyzes the installment payments, the unpaid balance, principal amounts paid per period, total installment payment, and current cumulative equity. It also calculates the "equity break-even point" (Badiru, 2016) for the debt being analyzed. The equity break-even point indicates the time when the unpaid balance on a loan is equal to the cumulative equity on the loan. With the output of this program, the basic cost of servicing the project debt can be evaluated quickly. A part of the output of the program presents the percentage of the installment payment going into equity and interest charge respectively. The computational procedure for analyzing project debt can be found in Badiru and Omitaomu (2007, 2011). The point at which the curves of the equity accumulated and the unpaid balance intersect is referred to as the *equity break-even point*. It indicates when the unpaid balance is exactly equal to the accumulated equity or the cumulative principal payment. The importance of the equity break-even point is that any equity accumulated after that point represents the amount of ownership or equity that the debtor is entitled to after the unpaid balance on the loan is settled with the project collateral. The implication of this is very important, particularly in the case of mortgage loans. The equity break-even point can be calculated directly from the formula derived for that purpose (see Badiru 2016).

Other pertinent technology cost analysis topics include technology cost estimation, capital allocation over competing technologies, capital rationing, and investment optimization. Coverage of these topics and other related topics can be found in the references for this chapter (Badiru and Omitaomu, 2007, 2011; Badiru, 1996, 2009, 2016; 2019).

CONCLUSION

In any Industry 4.0 enterprise, computations and analyses similar to those suggested in this chapter are crucial in controlling and enhancing the bottom-line survival of the organization. Analysts can adapt and extend the conventional or standard techniques for application to the prevailing scenarios in their respective organizations.

REFERENCES

Badiru, Adedeji B. (1996), *Project Management in Manufacturing and High Technology Operations*, 2nd Edition, John Wiley & Sons, New York, NY.

Badiru, Adedeji B. (2009), *STEP Project Management: Guide for Science, Technology, and Engineering Projects*, Taylor & Francis/CRC Press, Boca Raton, FL.

Badiru, Adedeji B. (2016), "Equity Breakeven Point: A Graphical and Tabulation Tool for Engineering Managers," *Engineering Management Journal 28*(4): 249–255.

Badiru, Adedeji B. (2019), *Project Management: Systems, Principles, and Applications*, 2nd Edition, Taylor & Francis/CRC Press, Boca Raton, FL.

Badiru, Adedeji B. and O. A. Omitaomu (2007), *Computational Economic Analysis for Engineering and Industry*, Taylor & Francis/CRC Press, Boca Raton, FL.

Badiru, Adedeji B. and O. A. Omitaomu (2011), *Handbook of Industrial Engineering Equations, Formulas, and Calculations*, Taylor & Francis/CRC Press, Boca Raton, FL.

Computational Tools for Systems 4.0

7

INTRODUCTION TO DIGITAL TWIN FOR SYSTEMS 4.0

The recent advances in digital transformation are metamorphosing Industry 4.0 for effective decision-making. In many applications, field assets, machines, products, plants, and factories are increasingly being connected to the Internet. This allows for these systems to be located, communicated with, analyzed, and controlled via the network. Thus, these digital transformations are revolutionizing new types of services and business models. On the computational side, cyber-physical systems (CPSs) have been proposed as a key concept of Industry 4.0 architecture. A CPS is a set of physical devices, equipment, and system that interact with virtual cyberspace through a communication network. The cyber-model of such a physical system is called a digital twin (DT). Simply stated, a DT is a digital replica of a physical system (or asset or process) in the built or natural environment. For actionable applications, such a replica must be a *realistic* and *dynamic* representation of the physical system. The word "twin" in DT implies that the replica system would be linked to the physical system throughout its entire life cycle. In other words, the digital replica can be treated as an entity on its own. The recent advances in smart sensors, Internet of Things, cloud computing, machine learning (ML), and artificial intelligence (AI) are enabling this transformation. Applying the DT concept to Systems 4.0 will allow the creation of fit-for-purpose digital representations of industrial operations and processes using collected data and information to enable analysis, decision-making, and control for a defined objective and scope (Kumbhar et al., 2023; Chaudhari et al., 2023; Stavropoulos and Mourtzis, 2022; Radanliev et al., 2022).

DOI: 10.1201/9781003312277-7

To achieve a realistic digital replica of a physical system, it is paramount to have good-quality data and well-managed data to achieve an effective and successful operation of a DT. Overall, the concept of DT should depend on the context and viewpoint required for a specific use case; that is, the digital twin is a fit-for-purpose digital representation. A DT need only collect the data relevant for the use case of interest rather than all available data from the physical system. This implies that there can be several DTs developed for a single physical system; these many DTs can also be integrated with one another to achieve a super DT for the physical system. Like all design processes, a good understanding of the scope and constraints of the intended DT is needed to avoid enormous errors and costs. Three key factors could help in assessing the scope and constraints of a DT (Grieves and Vickers, 2017):

1. Application – the specific application of the DT determines the rate of data intake and the fidelity needed to support the decision or control action that is desired. The scope and objectives of the DT decides which aspects should be included. The usefulness of a DT is determined by accuracy not complexity. Avoiding unnecessary components and functionalities saves time and effort and limits potential operational errors.
2. Viewpoint – the viewpoint of the desired decision or control action determines whether a product, process, or system twin is needed. Similarly, the viewpoint determines the methods and tools required, such as if the DT should be an emulator (mimic the observable behavior of the system) or a simulator (model the state of the system).
3. Context – the context needed to support a viewpoint determines how information should be provided by the DT. For example, visualization (while powerful) may not be needed on the requirements of the use case and constraints on time and budget. A DT includes both static and dynamic information. The static information includes geometric dimensions, processes, and so on. The dynamic information includes sensors data, real-time operational data, and other information that changes over time and during runtime. Similarly, the relevant viewpoint can influence the types of information that form the context provided by the DT.

Some of the several potential ways that a DT can be used in Systems 4.0 include:

1. Minimizing the impact of downtime – an infrastructure "health twin" can use process and equipment data to monitor, troubleshoot, diagnose, and predict faults and failures in infrastructure.

The knowledge can then inform the control of the infrastructure system or subsystems. Equipment downtime is a major liability in the manufacturing industry. Industrial processes often involve the use of multiple equipment simultaneously. To decide the best combination of equipment that would enhance manufacturing outputs, a DT can be used to determine the optimum combination in terms of costs, outputs, and manpower.

2. Optimizing infrastructure planning and scheduling – an "operative twin" can collect data from infrastructure sensors and operations to analyze the status of infrastructure operation and any fluctuations in resources, etc. This exploration should be useful for evaluating the impacts of extreme events on Systems 4.0. These benefits can be derived using DTs in the energy production industry. For example, a DT of physical asset can be used to perform various operations from the real-time data collected in the power grid industry to evaluate reliability and resilience metrics. DTs can also be used in maintenance and repairs of power plants and electrical substations.

3. Enabling virtual commissioning – a "commissioning twin" can use information from the vendor and data collected by monitoring new equipment performance during commissioning to enable system optimization and continuous improvement. This knowledge can allow infrastructure operators and owners to discover and resolve issues before investment and avoid the need for costly adjustments during or after installation. This has a potential benefit in the steel industry, where one of the major concerns in the manufacturing of steel products is rusting. A DT can be simulated to test the steel product under different weather conditions with varying alloy compositions to arrive at the most suitable composition.

4. Monitoring system performance – a "situational awareness twin" can use information from operational metrics to evaluate system resilience to external stresses. This assessment should help system operators to operate the system more effectively and efficiently. A situational awareness twin can be developed for monitoring road traffic patterns to explain the bottlenecks leading to suboptimal traffic flow along the transportation corridor during extreme weather events such as flooding and to design, test, and evaluate the appropriate mitigation strategies in cyber space before the real-world implementations.

Thus, a simulatable DT is a risk-free experimentation aid to explore and quantitatively evaluate the efficiency of various operational strategies, interventions, adaptation choices, and design alternatives for maximizing throughput.

DIGITAL TWIN DESIGN

Developing a DT needs different components, including sensors, communication networks, and a digital platform. The DT design and development process can be categorized into four phases using the DEJI™ model (Badiru, 2022): a) Design, b) Evaluate, c) Justify, d) Integrate.

1. The design phase demands that the physical systems and digital system coordinate and communicate in a single operation-oriented design of the system. Design phase will involve architecting the virtual representation of the physical system's hardware sub-components. Design phase can be performed on a physical computing platform or in the cloud. Activities in this phase include: (i) an accurate and adequate definition of the objectives and scope of the analysis; (ii) an identification of the critical elements of the assets to be modeled.

2. The evaluate phase pertains to understanding the performance and behavioral elements of the physical system. These include attributes such as system tolerance, stress, and design. This phase is related to the simulation aspects of the DT. Simulation will provide an estimate of DT's operations with respect to the physical system operations. Based on the simulated operations, system-level influencing factors and parameters can be determined. These controllable factors and parameters will be critical toward making performance and operational improvement decisions on the physical system. Some of the steps in this phase include: (i) the acquisition of static and dynamic data elements for the system; (ii) the processing of the data elements and their integration into a common data framework; and (iii) the construction of the detailed DT using an automated data processing pipeline.

3. The justify phase is basically about connecting the utility of the finished DT to the original objectives and scope using quantitative metrics. Some of the specific steps include: (i) detailed documentation of the business case for the DT and its alignment to the vision of the organization; (ii) quantifying the operational feasibility of the DT, this is crucial for the operational acceptability of the DT; and (iii) analyzing the benefit/cost implications of the DT.

4. The integrate phase corresponds to the actual operations of the physical system and its attributes such as system's age, operational constraints, etc. In the integrate phase, the DT is expected to run in real time alongside the physical system. The steps in the integrate

phase include: (i) the analysis of the DT using various models and algorithms; (ii) the evaluation of the overall reliability of the DT for the intended objectives; and (iii) the continuous improvement of the DT system for robust performance. The continuous improvement task includes improvement in data quality, quantity, and or availability, increase in scope and detail level, and modification to (and/or addition of new) performance indicators.

A standard framework can help enable systems engineers leverage DTs for decision-making and control by providing the means to navigate the complex set of standards, technologies, and procedures that can be used for implementation. The framework should be generic, reusable, and customizable irrespective of implementation so that it can support multiple relevant use cases. Such a framework should include guidelines, methods, and best practices. The framework can enable the generation and management of common data and model components to enhance the reuse of these components, e.g., a model components library or model templates may be useful for reusing and composing model components. A generic framework should include four parts (West and Blackburn, 2017):

- Overview and general principles
- Reference architecture
- Digital representation
- Information exchange

These four parts will provide guidelines and procedures for analyzing modeling requirements, defining scope and objectives, promoting common terminology usage, specifying a generic reference architecture that enables the instantiation of the digital twin for a specific use case, and supporting information modeling of the physical system and information synchronization between a DT and physical system.

CHALLENGES OF DEVELOPING AND USING DIGITAL TWINS

Since DT is about improving the availability of actionable information, the major challenge in developing a DT is the quality and quantity of available data for the development of the DT. The scope and level of detail required influence the overall data need. Too much detail makes the need for more data increasingly important, and, of course, more challenging. While the data collection methods have, over the years, undergone several evolutions, the multiplicity of databases or different data sources present an obstacle to information consolidation. Advances in digital transformations can also help with data collection by deploying sensors to collect new or better data. The data collection process, thus, involve significant investments in human and equipment resources, which is also a challenge for the development of a DT. Thus, this could impact the acceptability of a DT within the organization.

In addition to data availability and access, a DT requires real-time connection to and synchronization with the physical system. This is usually a very difficult proposition due to the requirements on the part of the workers. Since the virtual connection is crucial, there should be appropriate strategies in place to stop this requirement becoming a problem. One of those strategies is to start a DT as a stand-alone system using near real-time data ingestion during the evaluation of the reliability of the DT. A carefully managed virtual connection could also enhance the acceptance of the DT across the organization.

There is also concern about the cybersecurity vulnerability of connecting the DTs to the physical and hosting DTs in the cloud (Alcaraz and Lopez, 2022). DTs are vulnerable to cyber-attack physically and digitally. If not handled appropriately, cybersecurity could not only threaten the operational requirements of DTs, but also lead to mistrust in their deployments (Alcaraz and Lopez, 2022). Systems 4.0 approaches could also be used to address some of the security challenges of DTs.

BENEFITS OF DTs

In the past decade, DT technology has enabled organizations in multiple domains, including the steel industry, aerospace industry, healthcare industry, transportation industry, energy industry, among others, to explore its benefits. Among these benefits, irrespective of the size and scale of the DT, are (Stavropoulos and Mourtzis, 2022; Radanliev et al., 2022; Grieves and Vickers, 2017):

- Enabling system owners, operators, and decision-makers to simulate and model performance in varying scenarios
- Updating and upgrading physical items or assets
- Identifying and analyzing issues for improved system performance
- Predicting system outcomes with higher certainty
- Unlocking latent value from systems, processes, technologies, and people.

COGNITIVE DIGITAL TWINS FOR SMART SYSTEMS 4.0

The concept of cognitive DT (CDT) is an extension of DT (Ali et al., 2021). CDT incorporates three additional layers (Ali et al., 2021): access, analytics, and cognition. The idea of CDT brings the human dimension into DT (Agrawal et al., 2023). The *access layer* is the enhanced communication layer, especially with respective to access to data about the state of the physical system. The *analytics layer* brings advanced ML and AI into the framework to enhance actionable knowledge. The *cognitive layer* enables human cognition to convert the traditional DTs into smart and intelligent systems.

The concept of CDT is closely related to Human–Computer Interaction (HCI) and Human–Machine Interaction (HMI), which deals with achieving smooth interfaces between humans and DTs (Alcaraz and Lopez, 2022). Using the Triple C Model (Badiru, 2008) of project management—Communication, Cooperation, and Coordination—CDTs could alleviate some of the challenges associated with the interfaces and help achieve: (i) productive engagement through effective communication; (ii) collaborative task performance through optimal resource allocation and commitment; and (iii) trust in automation through transparency of the DT operations. To harness the full potential of CDT, the following questions must be resolved:

- What are the roles of DTs?
- What are the roles of humans?
- What approvals are needed to operationalize recommendations from DTs?
- What approvals are needed for humans to ignore the recommendations from DTs?
- What is the procedure for handling unintended consequences from DTs?

REFERENCES

Agrawal, A., Thiel, R., Jain, P., Singh, V., and Fischer, M. (2023), "Digital Twin: Where do humans fit in?" *arXiv preprint arXiv:2301.03040.*

Alcaraz, C., and Lopez, J. (2022), *Digital Twin: A Comprehensive Survey of Security Threats*, IEEE Communications Surveys & Tutorials.

Ali, M. I., Patel, P., Breslin, J. G., Harik, R., and Sheth, A. (2021), "Cognitive digital twins for smart manufacturing," *IEEE Intelligent Systems 36*(2): 96–100.

Badiru, A. B. (2008), *Triple C Model of Project Management: Communication, Cooperation, and Coordination*, CRC Press, Boca Raton, FL.

Badiru, A. B. (2022), *Systems Engineering Using the DEJI Systems Model®: Evaluation, Justification, and Integration with Case Studies and Applications.* CRC Press.

Chaudhari, P., Gangane, C., and Lahe, A. (2023), "Digital Twin in Industry 4.0 A Real-Time Virtual Replica of Objects Improves Digital Health Monitoring System," in Lalit Garg, Dilip Singh Sisodia, Nishtha Kesswani, Joseph G Vella, Imene Brigui, Peter Xuereb, Sanjay Misra, Deepak Singh (eds.), *International Conference on Information Systems and Management Science*, pp. 506–517, Springer, Cham.

Grieves, M., and Vickers, J. (2017), "Digital Twin: Mitigating Unpredictable, Undesirable Emergent Behavior in Complex Systems," in Franz-Josef Kahlen, Shannon Flumerfelt, Anabela Alves (eds), *Transdisciplinary Perspectives on Complex Systems*, pp. 85–113, Springer, Cham.

Kumbhar, M., Ng, A. H., and Bandaru, S. (2023), "A digital twin based framework for detection, diagnosis, and improvement of throughput bottlenecks," *Journal of manufacturing systems 66*: 92–106.

Radanliev, P., De Roure, D., Nicolescu, R., Huth, M., and Santos, O. (2022), "Digital twins: artificial intelligence and the IoT cyber-physical systems in industry 4.0," *International Journal of Intelligent Robotics and Applications 6*(1): 171–185.

Stavropoulos, P., and Mourtzis, D. (2022), "Digital twins in industry 4.0," in Dimitris Mourtzis (ed.), *Design and Operation of Production Networks for Mass Personalization in the Era of Cloud Technology*, pp. 277–316, Elsevier, Cambridge.

West, T. D., and Blackburn, M. (2017), "Is Digital Thread/Digital Twin Affordable? A Systemic Assessment of the Cost of DoD's Latest Manhattan Project," *Procedia Computer Science 114*: 47–56.

Index

Printed in the United States
by Baker & Taylor Publisher Services